THE FAST MISSION

THE FAST MISSION

Edited by

R. F. PFAFF, Jr

NASA / Goddard Space Flight Center, Greenbelt, Maryland, USA

Reprinted from *Space Science Reviews*, Volume 98, Nos. 1–2, 2001

KLUWER ACADEMIC PUBLISHERS

DORDRECHT / BOSTON / LONDON

A.C.I.P. Catalogue record for this book is available from the Library of Congress

ISBN: 0-7923-7046-5

Published by Kluwer Academic Publishers,
P.O. Box 17, 3300 AA Dordrecht, The Netherlands

Sold and distributed in North, Central and South America
by Kluwer Academic Publishers,
101 Philip Drive, Norwell, MA 02061, U.S.A.

In all other countries, sold and distributed
by Kluwer Academic Publishers,
P.O. Box 322, 3300 AH Dordrecht, The Netherlands

Printed on acid-free paper

Printed in the Netherlands

TABLE OF CONTENTS

AN OVERVIEW OF THE FAST AURORAL SNAPSHOT (FAST) SATELLITE

R. PFAFF[1], C. CARLSON[2], J. WATZIN[1], D. EVERETT[1] and T. GRUNER[1]
[1] *NASA/Goddard Space Flight Center, Greenbelt, MD, U.S.A.*
[2] *University of California, Berkeley, CA, U.S.A.*

Abstract. The FAST satellite is a highly sophisticated scientific satellite designed to carry out *in situ* measurements of acceleration physics and related plasma processes associated with the Earth's aurora. Initiated and conceptualized by scientists at the University of California at Berkeley, this satellite is the second of NASA's Small Explorer Satellite program designed to carry out small, highly focused, scientific investigations. FAST was launched on August 21, 1996 into a high inclination (83°) elliptical orbit with apogee and perigee altitudes of 4175 km and 350 km, respectively. The spacecraft design was tailored to take high-resolution data samples (or 'snapshots') only while it crosses the auroral zones, which are latitudinally narrow sectors that encircle the polar regions of the Earth. The scientific instruments include energetic electron and ion electrostatic analyzers, an energetic ion instrument that distinguishes ion mass, and vector DC and wave electric and magnetic field instruments. A state-of-the-art flight computer (or instrument data processing unit) includes programmable processors that trigger the burst data collection when interesting physical phenomena are encountered and stores these data in a 1 Gbit solid-state memory for telemetry to the Earth at later times. The spacecraft incorporates a light, efficient, and highly innovative design, which blends proven sub-system concepts with the overall scientific instrument and mission requirements. The result is a new breed of space physics mission that gathers unprecedented fields and particles observations that are continuous and uninterrupted by spin effects. In this and other ways, the FAST mission represents a dramatic advance over previous auroral satellites. This paper describes the overall FAST mission, including a discussion of the spacecraft design parameters and philosophy, the FAST orbit, instrument and data acquisition systems, and mission operations.

1. Introduction

Progress in understanding particle acceleration in the Earth's magnetosphere and related space physics processes has traditionally been led by advances in scientific instrumentation. Rather than just filling in details, high resolution measurements of auroral 'microphysics' have consistently unveiled important new mechanisms inherent to fundamental particle acceleration physics and related natural processes. NASA's Fast Auroral SnapshoT (FAST) satellite is a unique scientific observatory that explores the detailed physical processes of charged particle acceleration that create the visible aurora in the polar regions of the Earth. Its primary observations are high spatial and temporal resolution measurements of electric and magnetic fields and charged particles.

Space Science Reviews **98:** 1–32, 2001.

TABLE I
FAST science team

University of California, Berkeley	Dr Charles W. Carlson (Principal Investigator)
	Dr Robert E. Ergun
	Dr James P. McFadden
	Professor Forrest S. Mozer
	Dr Michael A. Temerin
University of Minnesota	Professor Cynthia A. Cattell
Los Alamos National Laboratory	Dr Richard C. Elphic
Lockheed Space Science Laboratory	Dr David M. Klumpar
	Dr William K. Peterson
	Dr Edward G. Shelley
University of New Hampshire	Professor Eberhard Moebius
University of California, Los Angeles	Dr Robert J. Strangeway
NASA/Goddard Space Flight Center	Dr Robert F. Pfaff (Project Scientist)

FAST is the second Small Explorer (SMEX) satellite selected by NASA to carry out rapid, low cost, and highly focused scientific investigations (Baker *et al.*, 1991). The philosophy of the SMEX program mandates that a single Principal Investigator (P.I.) be firmly in charge of all aspects of the mission. Such a program is ideally suited to a mission such as FAST. The Principal Investigator for the FAST mission is Dr. Charles W. Carlson of the Space Sciences Laboratory of the University of California at Berkeley. He is supported by a small team of co-investigators listed in Table I.

The FAST satellite was launched from the Western Test Range at the Vandenberg Air Force Base on August 21, 1996 by a Pegasus-XL vehicle that was released from the underside of an L-1011 jet aircraft at Mach 0.8 at 11.9 km altitude. Built by the Orbital Sciences Corporation, the three stage solid propellant system of the Pegasus-XL placed FAST in its high latitude (inclination 83°) elliptical orbit of 350 by 4175 km. The satellite crosses the auroral zones (which form ovals at ~65° magnetic latitude north and south) four times per orbit. The launch date was selected so that FAST's apogee and initial local time coverage (in the noon-midnight sector) would be ideally positioned to provide measurements during the northern hemisphere winter months when coordinated ground-based and optical auroral observations, including those from NASA's Polar satellite, would be optimum.

FAST represents dramatic improvements over previous NASA auroral missions. Not only do the FAST sensors and on-board flight computer acquire data several orders of magnitude faster than previous missions, but the orientation of the instruments with respect to the magnetic field provides the first continuous measurements of energetic particles at all pitch angles independent of spacecraft spin. The FAST science goals and instruments have benefited significantly from results provided by previous auroral missions, such as Freja, Viking, S3-3, and Dynamics Explorer. Furthermore, many of its instruments were developed over the years in NASA's sounding rocket program which provided not only a test bed for instrument development, but also significant high resolution measurements of auroral acceleration phenomena at low altitudes (500–1500 km) which helped formulate several of the science questions that FAST was designed to address.

The data collection strategy of FAST is based on the fact that auroral processes occur in very limited bands (typically 5–10° wide sectors that constitute the ovals of auroral light that surround each pole), and thus high time resolution measurements need not be taken throughout the orbit. Instead, the instruments are programmed to take 'snapshots' of the auroral acceleration phenomena (at rates as high as 80 Mbits s^{-1}), and then to telemeter these data to the ground at a slower rate at convenient times when ground-station receivers are both in view and available. Furthermore, the FAST on-board computer includes several different modes so that the sampling rates from each sensor may be tailored for a particular aspect of the investigation. Since both fields and particle data are collected within the auroral acceleration region, such simultaneous measurements enable cause and effect to be distinguished among competing auroral physics phenomena.

Both the FAST satellite and instrument designs required a highly integrated approach in order to minimize mass and power requirements, to optimize the instrument performance, and to maintain low cost. In this manner, FAST benefits significantly from the single P.I. and central processing approach. One tangible advantage of this philosophy is that a single instrument processing unit on the spacecraft controls all of the instrument operations, including their commanding and data acquisition and storage. This highly centralized unit enables significant mass and power savings to the FAST design while providing optimum efficiency for the instrument operation and triggering algorithms. Furthermore, this approach facilitates the creation of a single data set for all of the instruments that is provided for each scientist from a single source which is subsequently analyzed using a single set of software.

All of the FAST scientific instruments are working exceptionally well, returning high time resolution fields and particles observations. Although one spin plane electric field boom did not deploy completely, the electric field instrument is fully operational, since sensors subtended on three booms are sufficient to provide the spin plane electric field vector. In addition to the instruments, the other satellite subsystems (e.g., power, telemetry, commands, thermal control, and attitude) are also operating nominally. Although the orbit of FAST passes directly through the Earth's intense radiation belts, the FAST instruments, data system, and payload subsystem have shown little, if any, degradation since launch, with not one SEU (single event upset) recorded thus far. Not only has FAST operated through its planned two-year prime mission, it is anticipated that the satellite will continue to operate well past the current period of enhanced solar activity.

In accord with current NASA guidelines, the FAST mission has an open data policy. Survey data are made quickly available to the public (e.g., via the World Wide Web) for event identification and numerous studies. This survey data is, by itself, at a higher resolution than that of most previous NASA auroral missions.

An important aspect of the FAST Small Explorer mission is its linkage to scientific research carried out by several other scientific teams in the space physics community. Observing 'campaigns' have been carried out in which sounding rockets and dedicated ground-based and airborne all-sky cameras, auroral TV, and magnetometers are operated in conjunction with coincident FAST passes overhead. In addition, FAST provides an important low-altitude complement to the International Solar-Terrestrial Physics (ISTP) program which includes the NASA Polar Mission, which images the aurora from its vantage point near 9 Earth radii above the Earth's polar regions.

This paper serves as both an overview of FAST mission and as an introduction to the subsequent papers in this volume. We begin with a brief review of those aspects of auroral science that were particularly critical in designing the FAST mission, followed by an overview of the FAST spacecraft and instruments. We conclude with a discussion of the FAST orbit and mission operations. Detailed descriptions of the FAST scientific instruments, deployment systems, on-board flight computer, and mission operations appear in subsequent articles. Initial scientific results from the FAST mission are described in Carlson *et al.* (1998, and references therein), as well as in McFadden *et al.* (1999).

2. The Aurora as a Mission Design Parameter

An image of the Earth's northern polar region and auroral oval is shown in Figure 1, captured by the Wideband Imaging Camera (Mende *et al.*, 2000) on NASA's IMAGE satellite. This figure illustrates the location of auroras as extended intercontinental ovals centered on the north and south magnetic poles. Depending on solar activity, the aurora can be quiet and steady or very active and dynamic, in which case its thickness and its equatorward boundary will vary significantly. The FAST science objectives require the satellite measurements to be concentrated in the regions of the auroral ovals.

Since the Earth's auroral regions are narrow in latitude and are traversed at high speeds by a satellite at relatively low altitudes such as FAST, the instruments are programmed to acquire data at a very high rate only as the satellite traverses such regions. These collected data include 'bursts' or very high time resolution 'snapshots' that are evaluated by the on-board flight computer. Only events satisfying a pre-determined, programmable trigger criterion are stored at maximum resolution and later transmitted to the ground. The bursts are triggered by a variety of phenomena, depending on the science problem being studied. In this fashion, a trigger from one instrument is typically used to acquire high-resolution data from

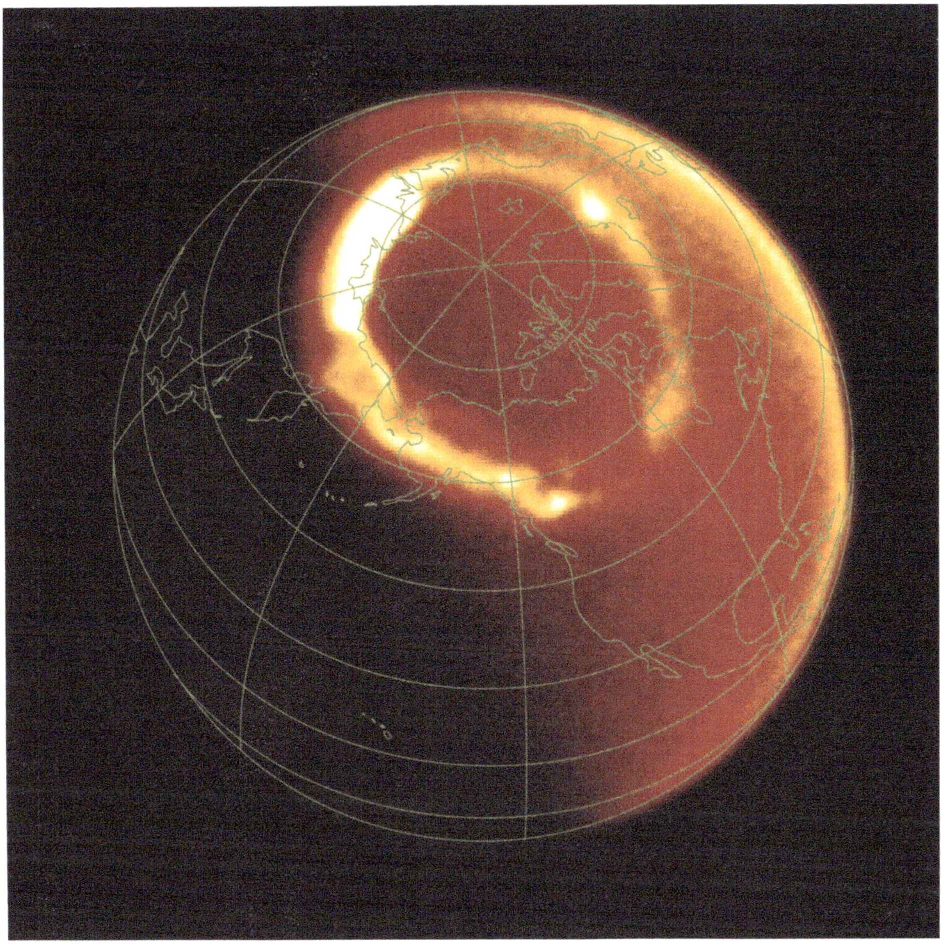

Figure 1. A global image of the Earth's polar region and aurora gathered with the Far Ultraviolet Imager on NASA's IMAGE Satellite [Courtesy, S. B. Mende, University of California, Berkeley].

all instruments. During special 'campaign' modes, data collection may be initiated by ground command in coordination with passes of other satellites and/or with airplane and ground correlative observations.

Although the light that produces the visible aurora is emitted in the upper atmosphere primarily at low altitudes (100–200 km), the acceleration of the particles that produces the aurora takes place at much higher altitudes, as shown in the sketch in Figure 2. This 'acceleration region' exists at altitudes of about 2000–10 000 km above the Earth in the high latitude regions. For this reason, the FAST satellite required an apogee as high as possible within this region, although for cost reasons it was limited to the performance confines of the Pegasus-XL launch vehicle. As a result, the satellite mass was made as light at possible and the satellite achieved

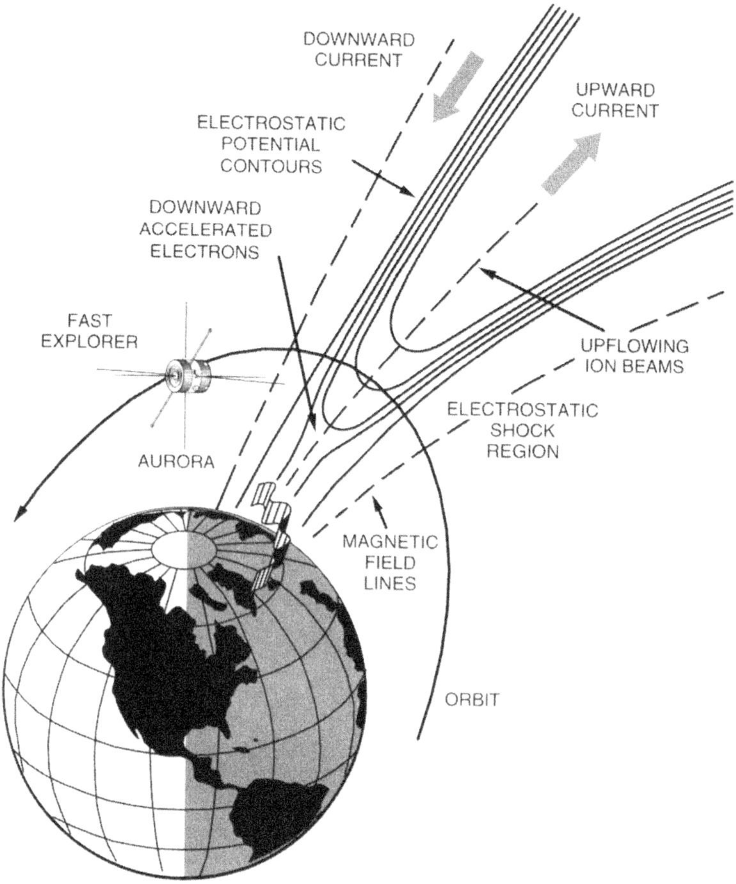

Figure 2. A sketch depicting some of the processes that occur in the Earth's auroral acceleration region as hypothesized at the time of the FAST launch. (Results from the FAST satellite have substantially revised this picture.)

an apogee of 4175 km, well above the originally proposed minimum altitude of 2500 km.

The primary scientific objectives of the FAST Mission are to understand how particles are accelerated to create the aurora and to investigate the microphysics of these and related space plasma processes. The processes that accelerate auroral particles along magnetic field lines and that sustain the auroras over time scales ranging from seconds to hours cannot be simple. It is for these reasons that comprehensive, high-resolution measurements of both fields and particles must be carried out by a single spacecraft positioned directly in the acceleration region itself.

As an example, most theories involve an electric field component parallel to (or along) the magnetic field direction to accelerate charged particles down, along the magnetic field lines, into the upper atmosphere. The magnetic field lines found in space around the earth, however, are such excellent conductors of electricity that conventional theories predict that they should promptly short circuit any 'parallel' electric field that might produce acceleration. Thus, such parallel electric fields would only exist in small spaces and for short times. In order to observe these fundamental, though short-lived, processes, the FAST instrumentation requires high temporal resolution.

The acceleration mechanisms that produce the aurora are also associated with a wide variety of other phenomena including upgoing beams of energetic ions and electrons, ion conics, intense electric field structures, and natural waves in space whose frequencies range from a few Hz to several MHz. The Earth's auroral acceleration processes are considerably complex, and many such phenomena undoubtedly result from a combination of electric fields, waves, and certain distributions of energetic ions and electrons. Indeed, a large part of the scientific question that the FAST mission addresses pertains to determining which processes are related to the cause, and which to the effect, of auroral acceleration phenomena.

The FAST Small Explorer thus provides an unprecedented opportunity to resolve the wealth of fine structure inside the auroral acceleration region and reveal the fundamental processes associated with such microphysics. Examples of science questions that the FAST data are being used to investigate include:

- how electrons and ions are accelerated in space to create the aurora;
- whether localized, parallel electric fields are set up to cause the acceleration and what are the various mechanisms that could sustain these fields, even briefly;
- which mechanisms efficiently heat and accelerate ionospheric ions (that subsequently populate the magnetosphere) and how do they work;
- how are radio waves and other emissions created during the acceleration processes; and
- how, and to what extent, are auroral acceleration processes tied to larger scale energy sources in the Earth's magnetosphere.

In order to achieve these science objectives, the FAST satellite carries highly specialized, state-of-the-art scientific instruments to the auroral acceleration regions to measure simultaneously the electric and magnetic fields, plasma waves (including their wavelengths and phase velocities), energetic electrons and ions including their full pitch angle distributions, energetic ion mass composition, and thermal plasma density and temperature. These science instruments were subsequently designed to function as an ensemble and in fact their data collection collectively triggers on a variety of auroral phenomena. The operation of the FAST instruments is orchestrated by a sophisticated on-board flight computer which was also designed and built by the FAST science team and hence is an essential part of the scientific success of the mission.

3. The FAST Spacecraft

3.1. General Design Approach

The philosophy of the FAST spacecraft design is based on the premise that the satellite basically consists of one integrated experiment that utilizes many different sensors. At all stages, the design of the ensemble of the instruments and spacecraft were optimized to achieve the highest quality measurements of auroral phenomena. The following spacecraft design guidelines were followed, insofar as possible:

(1) The mass of the spacecraft, instruments, and all flight components would be made as light as possible in order to achieve the highest possible apogee within the capabilities of the designated launch vehicle.

(2) The design and placement of all components would be such that the satellite moment of inertia would be optimized to maximize the length of the spin axis electric field booms.

(3) A single flight computer would control all of the instruments and their data acquisition, provide one common memory for burst and data storage, provide all regulated power, and house all of the electronics boards not directly needed at the sensor locations.

(4) A separate electronics system, the Mission Unique Electronics (MUE) would handle basic life-support functions for the spacecraft, such as attitude control, battery charge control, command ingest, and safing functions, leaving the instrument computer free to operate at its maximum capability.

(5) In order to optimize shielding against radiation, the instruments and flight components would be situated within the spacecraft where they 'make sense' (e.g., batteries on the outside, critical flight computer components on the inside).

(6) No solar paddles or extended solar arrays would be used, as they disrupt the *in situ* measurements, block energetic particle orbits, cause unwanted shadows, and create deleterious wake effects.

(7) Electrostatic and electromagnetic cleanliness would be a spacecraft priority. The solar array and all exposed surfaces must be conducting and kept at the same potential as the spacecraft internal ground.

(8) A large solid state memory (1 Gbit) would be included instead of tape recorders.

(9) The satellite would utilize a variety of downlink rates, for which the highest rate is 2.25 Mbps, and uplink commanding would be available at 2 kbps.

(10) The spacecraft would be single string, with practically no redundancy.

(11) A NASA 'Class C' Quality Assurance Program would be utilized.

(12) In order to detect flaws and verify system performance, the spacecraft would be vigorously tested with all components integrated.

The scientific goals of FAST can only be met by repeatedly sampling the high altitude charged particle auroral environment. In order to achieve the necessary high apogee, near-polar orbit, the FAST mission uses a unique, lightweight spacecraft developed by the Small Explorer (SMEX) Project at NASA/Goddard Space Flight Center. A system design was selected that uses proven design concepts and flight qualified or readily available hardware wherever possible.

The FAST spacecraft is a small, lightweight, orbit-normal spinner with multiple on-orbit deployable booms. The spacecraft provides structure, power, thermal control, telemetry and communication links, attitude control, and health monitoring support for the scientific instruments. A photograph of the FAST satellite is shown in Figure 3. The spacecraft characteristics are provided in Table II. The spacecraft mass is 191 kg including 65 kg of instruments (with IDPU) and carries no on-board propulsion. The masses of the various components of the FAST satellite are provided in Table III. A drawing of the FAST spacecraft showing all of its booms deployed is shown in Figure 4.

3.1.1. *Structure*

The FAST primary structure is constructed of aluminum and includes a single deck on which the instruments and electronic boxes are mounted as shown in the exploded drawing in Figure 5. There are two magnetometer booms located 180° relative to each other in the spin plane, which are stowed along the spin axis for launch. In addition, there are two axial stacer electric field booms and four radial, equally spaced, wire electric field booms that are deployed after launch.

The instrument deck layout is shown in Figure 6. Placing all of the instruments and main spacecraft components on a single deck whose plane includes the spin plane electric field wire booms and deployed magnetometer booms is an important design feature of the FAST spacecraft. Such a single plane design, together with the placement of the instruments at the furthest edges of this plane, optimizes the spacecraft moment of inertia such that the spacecraft is thus able to support the longest possible spin axis electric field booms (see Pankow *et al.*, this issue). Notice also in Figure 6 that the design enables all connectors to be exposed and accessible for testing purposes when all of the instruments and the MUE are installed on the single deck.

Above and below the instrument deck, the solar array attaches to an essentially hollow 'shell' whose main purpose is to maximize the solar array area within the limits of the Pegasus shroud. The relatively low mass of the solar array shell has minimum impact to the satellite moment of inertia. Drawings of the side and elevation views of the FAST satellite within the Pegasus XL shroud are provided in Figures 7(a) and 7(b). The result is a satellite whose shape is optimized for power (e.g., solar cell area) and whose moment of inertia is optimized to support the longest possible spin axis electric field booms.

3.1.2. *Power*

The FAST power system is a direct energy transfer system. Excess charge is dissipated in shunts to prevent battery overcharge. Shunt driver boxes external to the Mission Unique Electronics (MUE) are provided to switch the excess solar array

Figure 3. A photograph of the FAST Small Explorer Satellite in the clean room.

TABLE II
Characteristics of the FAST Satellite

Launch date	August 21, 1996
Launch vehicle	Pegasus-XL
Orbit	
–Inclination	83°
–Apogee altitude	4175 km
–Perigee Altitude	350 km
–Orbital period	133 min
Size	
–Diameter	1.02 m
–Height	0.93 m
Satellite Mass	
–Total	191 kg
–Instruments (with IDPU)	65 kg
Telemetry	
–High (science)	2.25/1.5/0.9 Mbps
–Low (s/c 'health' only)	4 kbps
–Commands (uplink)	2 kbps
On-board data storage	
–Solid state memory	125 Mbyte (1 Gbit)
Power	
–Solar cells on s/c body	2.6 m^2, single junction GaAs
–Peak spacecraft power (all systems in use)	117 W
–Spacecraft power (orbit averaged)	52 W
–Instrument power, including IDPU (peak)	39 W
–Instrument power, including IDPU (orbit averaged)	19 W
–Transmitter power when in use	28 W
–Battery	9 A-hr 'Super' Nickel Cadmium
–Solar Cell Surface	Conducting, Indium Tin Oxide on cover glass
Attitude	
–Spin rate	12 rpm (closed loop spin control)
–Spin plane orientation	Within 10° of magnetic field direction in auroral zones (somewhat worse in So. Hemis.)
–Actuators	Magnetic torquers (2)
–Attitude knowledge	0.1° (typical processed accuracy)
–Attitude sensors	Sun sensor, horizon crossing indicators, 3-axis magnetometer
Nominal Mission Lifetime	1 year

TABLE III
FAST mass summary

	Mass (kg)
Instruments	
Electric field booms and sensors	
Axial boom/sensor units (2)	4.3
Wire boom/dual sensor units (4)	13.5
Magnetic field boom/sensor units (2)	5.3
TEAMS	7.7
ESA assemblies (4 stacks of 4 180° FOV sensors)	18.0
IDPU (including instrument electronics)	16.4
	65.3 kg
Spacecraft Bus	
MUE	14.7
Power system	
Battery and shunt box	12.5
Solar array	35.4
ACS	9.7
RF system	5.2
Thermal system	3.0
Harness	9.1
Test connector panel	0.4
Structure	26.5
Balance weight	5.3
Pegasus adapter (23″)	2.0
Miscellaneous	2.2
	126.0 kg
Total (FAST satellite)	191.3 kg

current to the shunts. The battery on the FAST spacecraft is a 9 Ampère-hour super NiCad battery. The FAST power system is described in detail by Schnurr *et al.* (1995).

The MUE power circuitry provides voltage/temperature (V/T) control, current control, over-voltage control and a precision current monitor that is used as the sensor for the software Amp-Hour integrator. The battery charge control circuitry holds eight relays which are used to enable and configure the state of the controller. All relays are magnetic latching to hold states in between modes and commanded configuration changes. If the voltage is out of tolerance for three cycles,

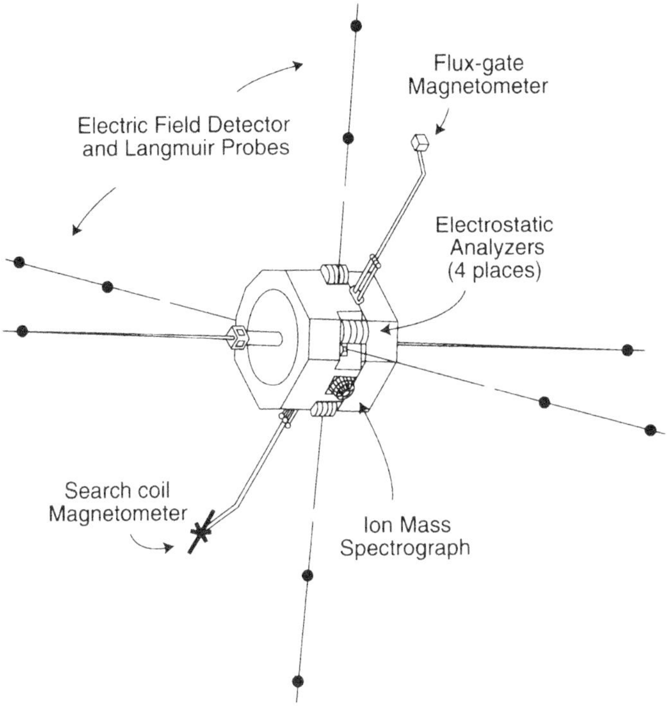

Figure 4. A drawing of the FAST satellite in its nominal flight configuration showing the nominal deployment of all of the booms. (In flight, one of the spin-plane electric field booms did not fully deploy.)

the MUE software will shut down non-essential loads. The charge control card drives the shunt regulator. The voltage and temperature controls are monitored and controlled according to pre-determined voltage-temperature (V/T) levels. In V/T control mode, when the battery reaches the selected voltage limit, the V/T controller will activate the shunt regulator. The battery temperature is read from platinum wire temperature sensors located in the battery. Sixteen different V/T levels are available and commandable from the ground. The overvoltage controller activates the shunt regulators when the battery approaches 34.5 ± 0.5 VDC. The power distribution function provides 11 relays to apply unregulated +31.5 V power to the instruments, deployment mechanisms, heaters, and ACS sensors. Bus power to all devices except the transponder is switched by the MUE. The transponder cycles power to the transmitter section only when the transmitter is in use.

3.1.3. *Solar Array*

A body-mounted solar array was required by the FAST science instruments in order to minimize plasma disturbances (both for field and particle measurements) around the spacecraft. Furthermore, to minimize electrostatic charging and stray magnetic fields, the mission required a conducting solar array (i.e., sealed inside a Faraday

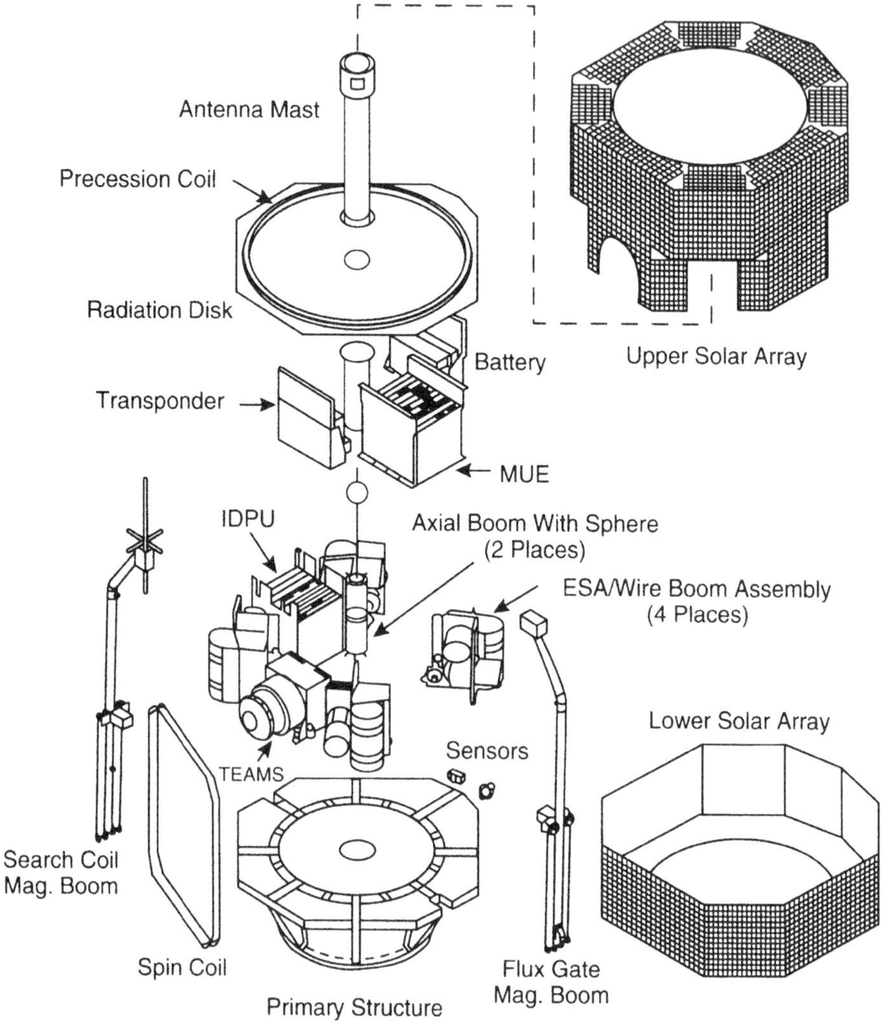

Figure 5. A drawing with an expanded view of the FAST satellite showing its major components.

box) with circuitry to cancel magnetic fields generated by the solar array strings and wiring.

The FAST solar array uses 8225 solar cells that comprise a total area of 2.58 m^2. The cells are 0.14 mm thick, 18.5% efficient single junction GaAs/Ge. A 60-mil thick coverglass was included to protect the cells from the harsh radiation environment. This coverglass is composed of fused silica with the standard space qualified anti-reflective front surface and ultraviolet reflective rear surface coatings and an additional outer surface transparent conductive coating of indium tin oxide (ITO) to act as part of the Faraday box. Details of the FAST solar array including a discussion of the solar cells with connection 'V-clips', boom shadowing, and other features of the array design are discussed in Kruer and Lyons (1994).

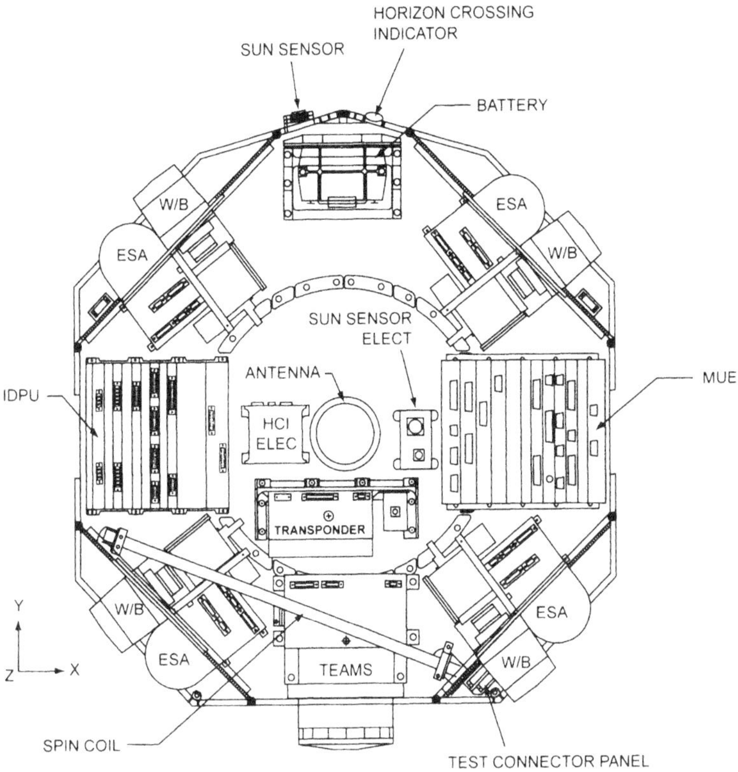

Figure 6. A drawing of the FAST instrument deck.

During its first 3.5 years in space, the FAST solar array output has varied from 60–130 W depending on beta angle (the angle between the orbital plane and the sun-earth direction), as shown in Figure 8(a) (Lyons, personal communication, 2000). Figure 8(b) shows the projected area of the solar array averaged over a spin period for various beta angles. Notice that the measured solar array power closely tracks this projected area. The available solar array power degrades with time, at a rate of about 10% over 3 years, as shown in Figure 8(a). Throughout this period, the average spacecraft bus voltage has been maintained near +31.4 V. The excellent solar array performance is somewhat better than predicted as the FAST orbit exposes the satellite to an intense radiation environment combined with high operating temperatures and significant atomic oxygen flux at perigee.

3.1.4. *Thermal*

Temperatures are controlled primarily by passive thermal control elements. As described in Parrish (1999), only the battery has a dedicated radiator and thermostatically controlled heaters. (The transponder uses heaters during safehold operations.) All of the main electronics are heat sunk to the main deck. The body mounted solar

(a)

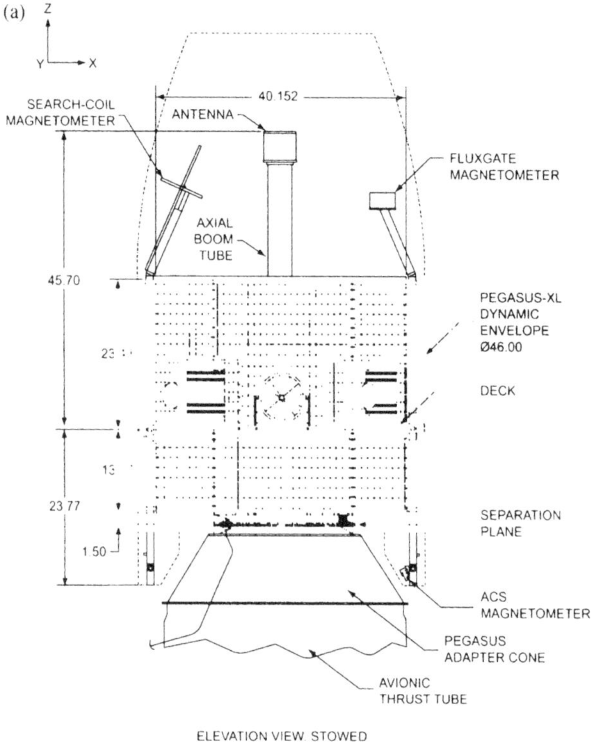

(b)

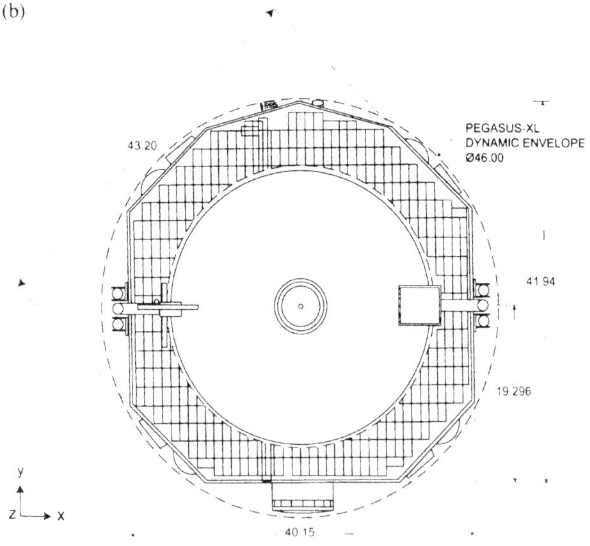

Figure 7. Drawings of the FAST spacecraft within the Pegasus-XL shroud, shown as an elevation or side view (a) and as a top view (b).

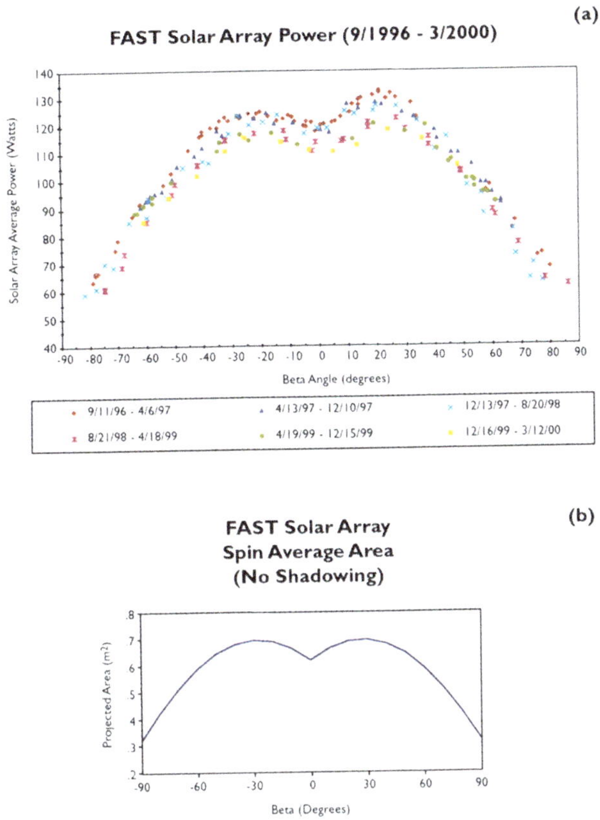

Figure 8. FAST solar array performance (a) and projected, spin-averaged solar array cell area (b).

array is conductively and radiatively coupled to the equipment deck and equipment cavity. The equipment power is transferred to the body array where it is then radiated to space. When the sun vector is nearly perpendicular to the spacecraft's spin axis, the body array provides a nearly room temperature environment for the electronic equipment.

Multi-layer insulation is positioned on the interior of select areas of the array to trim and adjust the equipment deck temperatures. Heat is also transferred to and from the thrust cone which supports the equipment deck. The thrust cone attaches to the launch vehicle irridited aluminum interface ring which protrudes from the bottom of the spacecraft.

During its first three years, the FAST internal instrument disk temperatures varied between −10 and 35 °C. The spacecraft has, in general, been somewhat warmer than predicted, but is still within temperature limits, as discussed in detail by Parrish (1999). In fact, the warmer spacecraft allowed data to be gathered during times when it was originally believed that sun-avoidance attitude maneuvers would be needed to avoid the cold temperatures that were predicted prior to launch.

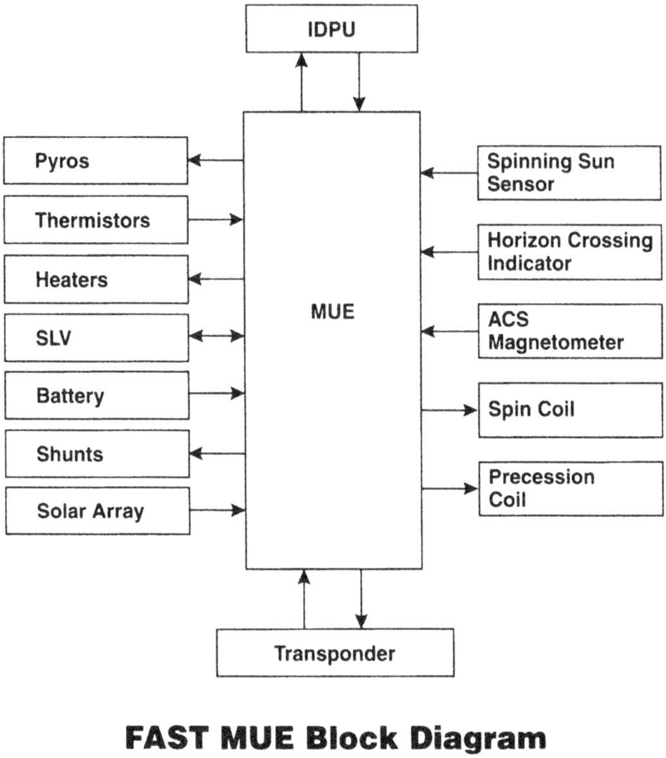

Figure 9. Block diagram of the Mission Unique Electronics (MUE).

3.1.5. *Mission Unique Electronics*

The spacecraft command and data handling (C&DH) system is embedded within the spacecraft electronics module. This module is known as the Mission Unique Electronics (MUE), and uses a pair of 2 MHz 80C85 8-bit microprocessors, with 72 kbyte ROM and 320 kbyte RAM. A block diagram of the MUE is shown in Figure 9. The MUE performs telecommand reception, stored command processing, telemetry data collection and generation, data encoding and decoding attitude control, power management and battery charge control, launch vehicle interface information and basic spacecraft health and safety functions. Also within the MUE is an 800 kbyte RAM recorder for capturing spacecraft safing events.

3.1.6. *Spacecraft (MUE) Software*

The FAST spacecraft (MUE) software is partitioned into three main subsystems: C&DH, ACS, and Power Management. A custom Operating System (OS) provides software task control and basic Input/Output (I/O) functions to the subsystems. The software processing is divided between the two 8085 microprocessors, with most C&DH functions in one, and the Power and ACS subsystems in the other. All of the software subsystems operate under three modes: Launch, Initial Acquisition, and Normal modes. The spacecraft was switched to normal mode after initial acquisition, where it will remain for the duration of the mission.

The C&DH command uplink system processes CCSDS formatted ground or stored commands and distributes them to other subsystems including the IDPU. The command system can process two types of stored commands: Absolute Time Commands (ATC), and Relative Time Commands (RTC). ATC's are stored in either of two buffers holding up to 512 commands each. These buffers hold commands to control spacecraft and some instrument operations that must be performed on a fixed timeline. RTC's can be stored in any of 64 sequence buffers of 13 commands each. Once commanded to start, these commands execute at times relative to each other. Up to 16 of these RTC buffers may be active at any time.

The power management subsystem software consists of two sections: the command processor and the Amp-Hour Integrator; both executing at 1 Hz. The Amp-Hour Integrator computes the battery state of charge based on the battery current and temperature. When the charge reaches 100%, the software issues a trickle charge command. The state of charge variable and several other variables computed by the power subsystem are also used by ground operators and safing checks for spacecraft power management. The command processing section interprets and executes ground commands to control spacecraft power distribution.

3.1.7. *Instrument Data Processing Unit*

The instrument data processing unit (IDPU), described briefly in Section 4.5 and in more detail in Harvey *et al.* (this issue), manages and controls the multiple functions of the fields and particles instruments, including boom deployments. Within the IDPU is a high density, 1 Gbit solid state recorder, which includes a 1–2 Mbyte partition for spacecraft health and safety data. The MUE and IDPU communicate via a simple serial interface at 9600 baud, with a high rate interface for the IDPU to directly access the transponder for downlink of science data. The IDPU contains a 32-bit microprocessor.

3.1.8. *Attitude Control System*

The primary objectives of the Attitude Control System (ACS) are to provide autonomous spin and precession control following separation from the launch vehicle, to control spacecraft spin and precession during the normal mode operations to meet the science imposed attitude requirements, and to maintain a spin rate and spin axis attitude consistent with the power and thermal requirements. The ACS is also responsible for processing ACS related ground commands and providing telemetry associated with the ACS.

The ACS for FAST is designed to maintain the spacecraft attitude as a simple spinner with a rotation rate of approximately 12 rpm. Pointing requirements are met by utilizing a complement of sensors, torquers and standard electronics in the MUE. Spin rate and spin axis orientation are maintained using two magnetic torquer coils. Either one spinning sun sensor, one horizon crossing indicator, or the spacecraft magnetometer can be used to measure the actual spin period. In addition to the active control elements, a fluid ring damper provides passive nutation control. The ACS provides closed-loop spin-rate control. Spin-axis precession is performed open loop and is closed via ground commands. The near orbit-normal spinner uses electromagnets to keep up with the daily orbit precession and to perform sun angle avoidance maneuvers to maintain the sun angle to less than 60°, if necessary.

3.1.9. *Attitude Knowledge*

FAST attitude knowledge consists of two elements: spin axis pointing and spin phase. Spin-axis pointing is verified by orbit detrending between the model magnetic field and the observed magnetic field. Nominal accuracy using actual FAST data is typically 0.1°, although this is a factor of two worse on orbits around torquer coil operations.

The spin-phase error is much less than 0.1 deg in sunlight (Sun-sensor determined), but jitter of this order is present when FAST is in eclipse (where the equivalent sun-phase is determined from using nadir-phase information). There are also frequent phase skips at eclipse entry and exit. In other words, although the spin period may be reasonably accurate in eclipse, the absolute phase can have problems.

An additional source of attitude error derives from spacecraft nutation. This error can be quite large, $\sim 0.2°$ with a 30-s period. Such effects are largest immediately after torquer operations. After computation on the ground, the FAST attitude knowledge is accurate to within approximately 0.1 deg.

3.1.10. *Telemetry Rates and Transmitter*

FAST uses a standard 5-W NASA transponder that receives commands at a data rate of 2 kbps that are transmitted as non-return to zero (NRZ) bi-phase modulation on a 16 kHz sub-carrier at 2.03964 GHz. Telemetry data is transmitted at 4 digital data rates (4 kbps, 900 kbps, 1.5 Mbps, and 2.25 Mbps) using NRZ phase modulation directly on the carrier. A multi-element micropatch antenna mounted on a boom above the spacecraft is used to support ground communications. FAST operations are discussed later in Section 5.

3.1.11. *Radiation*

The high radiation environment of FAST demanded careful part selection and careful attention to the structure in the design phase. All electronic components are supported by a machined aluminum equipment deck with an aluminum honeycomb radiation shield cover above. The solar array substrate is also honeycomb and provides radiation shielding on the sides. The solar cells are covered with extra-thick coverglass to help reduce the total-dose of radiation seen by the solar cells. Samples from each lot of parts were tested for single event effects and total dose effects. One reason why 8085 processors were selected for the MUE and the 32C016 processor was selected for the IDPU is that these were available as radiation-hard parts.

3.1.12. *Electric and Magnetic Shielding*

In order to minimize plasma disturbance around the spacecraft, the instruments require a solar array with magnetic and electrostatic cleanliness values of 7.85 nT and 0.1 V maximum voltage differential across the spacecraft outer surface, both an order of magnitude less than required for any array previously built by NASA. Throughout the design and fabrication phase, care was taken to reduce the electric and magnetic fields produced by the spacecraft so that the scientific measurements of these geophysical quantities would not be corrupted. To this end, the FAST solar array incorporates an integral Faraday cage. Each coverglass is coated with conductive (and transparent) indium tin oxide (ITO). The gaps between coverglass are filled with 'V'-shaped pieces of metal which are attached to the coverglass with palladium-filled epoxy. The edges of the solar arrays are covered with conductive foil. All external conductive surfaces are electrically connected together and connected to the aluminum structure, presenting an equipotential surface to the space plasma environment. The cells and backwiring were laid out to minimize magnetic fields. All harnessing which carries appreciable current is twisted to its return. Where redundant power lines were required, quad-and hex-twisted shielded lines were used so that all current would be close to its return. Power distribution relays were oriented to cancel internal magnetic fields. A loop was added to the battery harness to cancel magnetic fields produced within the battery. The spacecraft produces no more than 1 nT of DC magnetic fluctuations at the fluxgate magnetometer, and the search coil magnetometer detects negligible AC fields from the spacecraft (even before the magnetometer booms were deployed).

4. Overview of the FAST Scientific Instruments

The FAST design philosophy of a single-string spacecraft with little or no redundancy is reflected in the meticulous concern and attention to detail that the individual science investigators provided for their instruments. Building on a solid heritage from past satellites and sounding rockets, the scientific instruments were each uniquely designed for the FAST satellite and constitute integral parts of the spacecraft design and structure. A critical feature of this instrumentation is the highly sophisticated flight computer designed and built by the scientists at the University of California at Berkeley. This IDPU includes programmable trigger algorithms, burst memories, and a common 1 Gbit solid state data system used by all the instruments.

The FAST science instrument complement includes: a 3-axis, vector set of electric field/Langmuir double probes (extending 58 m tip-to-tip in the spin plane) which include additional inner sensors for wavelength and phase velocity measurements, vector fluxgate and search coil magnetometers for DC and AC magnetic field measurements, an ion mass spectrograph, an electron spectrograph, and electron and ion spectrometers. Control and management of these instrument components (including boom deployments) is performed by the IDPU.

The scientific instruments on the FAST satellite are described in detail in the accompanying papers in this volume. In this section, we present a brief overview description of each instrument. A list of the scientific instruments and their measurement characteristics and resolutions is provided in Table IV (a, b). In Figure 4, the locations of the scientific instruments on the FAST satellite are shown.

4.1. Electrostatic Analyzers (ESAs)

Quadrispherical electrostatic analyzers (ESAs) are used to measure electron and ion distribution functions. Particles enter the analyzer over a 360° field of view (FOV) and are selected in energy by the electrostatic analyzer, and imaged by micro channel plate (MCP) and discrete anodes. The 360° FOV is in the spacecraft spin plane which is aligned to within 6° of the magnetic field direction when the spacecraft is in the auroral zones. The out-of-plane FOV is 10° (5.5° at full-width half maximum (FWHM)) and 12° (7° FWHM) for the electron and ion sensors, respectively.

For the FAST satellite, the sensor heads are split into pairs of 'half-analyzers' located on opposite sides of the spacecraft. Sixteen half-analyzers are arranged into four ESA stacks located at 90° intervals around the spacecraft. Each ESA stack contains three Stepped ESA (SESA) analyzers that are used to gather the high time resolution electron measurements at specific energies, and a single ion or electron spectrometer (IESA or EESA) used to make detailed, full pitch-angle distribution measurements. Each analyzer in the stack can be commanded to operate in a variety of combinations of modes and configurations. The ESA stacks also include preamplifier-counter boards, an ESA logic-interface board, and high and low voltage supplies. The ESAs are described in detail by Carlson *et al.* (this issue).

4.2. Time-of-flight Energy Mass Angle Spectrograph (TEAMS)

The Time-of-flight Energy Angle Mass Spectrograph (TEAMS) instrument for the FAST payload is a high sensitivity, mass-resolving ion spectrometer with an instantaneous 360° × 8° field of view. TEAMS is designed to measure the full 3-dimensional distribution function of the major ion species (including H^+, He^+, He^{++}, O^+, O_2^+ and NO^+) within one half spin period of the spacecraft. The sensor energy range is between 1.2 and 12000 eV charge^{-1} and thus covers the core of all important plasma distributions in the auroral acceleration region.

The TEAMS instrument combines the selection of incoming ions according to their energy per charge by electrostatic deflection in a toroidal section analyzer with post-acceleration of up to 25 keV e^{-1} and subsequent time-of-flight analysis. The energy-per-charge analyzer is of the symmetrical quadrispherical type with a uniform response over 360° of polar angle (similar to the IESAs described above).

TABLE IVa
FAST particle detectors

Measurement	Coverage	Energy range	$\Delta E/E$	Sampling resolution	Field of view	Angular resolution	Sample array	Geometric factor $cm^2\ sr^{-1}$ E
Ion mass spectrometer	3D	2 eV – 10 keV	0.13	2.50 s (3D)	$360^\circ \times 10^\circ$	$10^\circ \times 22.5^\circ$	$4M \times 48E \times 64\Omega$ (3D)	0.015
				78 ms (2D)			$4M \times 48E \times 16\alpha$ (2D)	
Ion spectrometer	2D	3 eV – 24 keV	0.20	78 ms	$360^\circ \times 12^\circ$*	$11.2^\circ \times 12^\circ$	$48E \times 32\alpha$	0.009
Electron spectrometer	2D	4 eV – 30 keV	0.15	78 ms	$360^\circ \times 10^\circ$*	$11.2^\circ \times 10^\circ$	$48E \times 32\alpha$	0.005
Electron spectrograph	2D	4 eV – 30 keV	0.15	1.6 ms	$360^\circ \times 10^\circ$	$22.5^\circ \times 10^\circ$	$6E \times 16\alpha$**	$6 \times .01$

*Field of view can be deflected ± 10 deg from the center angle.
**A larger energy sample array can be achieved with poorer time resolution.

TABLE IVb
FAST field instruments

Measurement	Components	Frequency range	Sampling resolution	Measurement range	Measurement resolution	Max bit rate (kbits s^{-1})
Electric field						
DC E-field	3-axis	0–300 Hz	30 μs	± 1.6 V m^{-1}	16 bit	1600
Wave E-field	3-axis	300 Hz–16 kHz	30 μs	± 200 mV m^{-1*}	16 bit	1600
Swept frequency**	3-axis (E)	0–2 MHz	32 ms	(***) 10^{-15}–10^{-5} $(V/M)^2$/Hz	8 bit	0.75
receiver	1-axis (B)	0–2 MHz	32 ms	10^{-12}–10^{-5} nT^2/Hz	8 bit	0.50
E-field rectifier/filters	3-axis	100 kHz-2 MHz†	30 μs	0.1 mV m^{-1}–1 V $m^{-1\ddagger}$	8 bit	800
High speed burst	3E	0–1000 kHz	0.5 μs	± 200 mV m^{-1*}	10 bit	80 000
memory	1B	0–500 kHz	0.5 μs	± 1 nT		(4 chan.)
Plasma density	2 point	0–250 Hz	0.5 ms	(***) 1–6 10^4 nA	8 bit	32 (2 chan.)
Plasma temperature	–	–	1 s	0.1 eV–1 keV	8 bit	0.008
Magnetic Field						
Fluxgate magnetometer	3-axis	0–50 Hz	2 ms	$\pm 6 \times 10^4$ nT	16 bit	24
Search coil	2-axis	50 Hz–4 kHz	30 μs	(***) 10^{-1}–10^4 nT	16 bit	1600 (3 chan.)
magnetometer	1-axis	10 Hz–16 kHz	30 μs	4×10^{-2}–4×10^3 nT	16 bit	

*The amplitude range increases for short wavelength (less than boom length) waves up to 1.6 V m^{-1} peak.
** Frequency resolution, 15 kHz.
***Gain selectable.
† Selectable broadband range.
‡ Selectable gain, range varies with antenna length.

The full angular range of the analyzer is divided into 16 channels of 22.5° each. The TEAMS instrument is described in detail by Klumpar *et al.* (this issue).

4.3. MAGNETIC FIELD INSTRUMENT

The FAST magnetic field instrumentation includes both a vector DC fluxgate magnetometer and a vector AC search coil magnetometer. The fluxgate magnetometer is a three-axis instrument that uses highly stable low noise ring core sensors. The sensors are boom-mounted at two meters from the spacecraft body in a shielded housing. See Pankow *et al.* (this issue) for a discussion of the magnetometer boom systems.

The flux-gate sensor electronics provides drive signals for the sensors and amplifies and detects the second harmonic signals that are proportional to the magnetic field. The electronics digitizes the magnetic field information from DC to 100 Hz using a 16-bit analog-to-digital converter (ADC).

The search coil magnetometer uses a three axis sensor system that contains laminated permalloy cores, windings and preamplifiers. The system provides AC magnetic field data over the frequency range 10 Hz to 2.5 kHz on two axes. The third axis extends this response to 500 kHz. The electronics further amplify the signals prior to digitization. The FAST magnetometer sensors are described by Elphic *et al.* (this issue). The magnetic field data acquisition system is described in Ergun *et al.* (this issue).

4.4. ELECTRIC FIELD/LANGMUIR PROBE

The FAST electric field instrument was designed to deploy ten spherical sensors, two each on four 28 m, radial wire booms and one each on two axial stacers (see Figure 4). The spheres on the wire booms are located 28 m and 23 m from the spacecraft. The axial spheres are separated by 8 m tip-to-tip. Each sphere houses a preamplifier circuit. Unfortunately, one of the wire booms did not deploy properly. However, this did not preclude gathering vector electric fields by the FAST spacecraft.

Electric field components are derived from the voltage difference between two spheres that are in voltage mode. In current mode, a sphere is biased to a fixed voltage and the preamplifier output represents the current in the sphere (as a Langmuir probe). The fields electronics includes several data acquisition modes, including: low-frequency signal processing, high-frequency signal processing, swept frequency analyzer, plasma wave tracker, broad band filters, wave-particle correlator, high speed burst memory, high-frequency phase difference measurement, and fields digital signal processor. The FAST electric field/Langmuir probe instrument and fields data acquisition electronics are described in detail by Ergun *et al.* (this issue). The electric field boom systems are described by Pankow *et al.* (this issue).

4.5. Instrument Data Processing Unit (IDPU)

The IDPU uses a 10 MHz processor (32C016 derivative), with 16 kbyte ROM, 64 kbyte EEPROM, and 256 kbyte RAM for processor operations. Internal to the IDPU is a formatter card, implemented using programmable gate arrays, which is the primary hub for instrument control, data distribution, and timing. The formatter provides interfaces to the individual instrument electronics and performs high speed data acquisition, compression, averaging, and packetization of the science data. Also within the IDPU is a high density 1 Gbit (128 Mbyte) mass memory (solid state recorder) used for storage of collected science data. The mass memory includes error detection and correction (EDAC) circuitry to prevent corruption from radiation effects. (The recorder includes a selectable 1–2 Mbyte partition for spacecraft health and safety data.) The formatter arbitrates and sequences data into and out of the mass memory, and provides a DMA interface to the processor. Using ground-programmable table loads and commands, the IDPU flight software manages science telemetry data acquisition and performs necessary instrument control and housekeeping functions.

A central data system contains 1 Gbit of solid state memory which is partioned to provide continuous coverage of the auroral crossings at 'survey' data rates (0.5 Mbit s^{-1}) and to allow high speed (8–80 Mbit s^{-1}) 'burst' snapshots to be taken during the most interesting events. The fully integrated design was not only cost effective, but also produces a single data set with all the data available to each scientist for complete event analysis. The FAST Instrument Data Processing Unit is described by Harvey *et al.* (this issue).

5. FAST Orbit, Mission Operations, and Data Acquisition

FAST was launched on a Pegasus rocket on August 21, 1996, into a 350 by 4175 km orbit, inclined 83°. The orbit's alignment of perigee changes at −1.75° per day, and the right ascension of its ascending node changes at −0.5° per day. The orbital period is 133 min. Apogee traverses the acceleration regions every 3.3 months and complete local time coverage is obtained every 8.1 months, as shown in Figure 10. As these two periods beat together, comprehensive latitude and local time coverage can be obtained within a multi-year mission. Such extended observations provides the coverage needed to perform in-depth statistical analyses of numerous auroral processes.

FAST auroral science investigation requirements include a campaign style of mission operations while the FAST apogee is in the Northern Hemisphere during the local winter months. For the first campaign period, one to two passes of telemetry (acquired at Poker Flat, Alaska and Kiruna, Sweden) were scheduled every orbit (10–11 per day), with commands being sent at least once per day. (Note: this commanding is possible every orbit from Poker Flat during campaigns.) For the

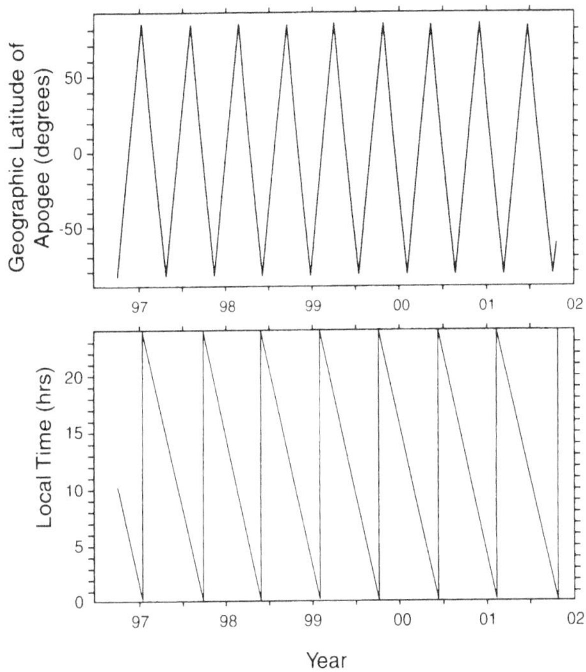

Figure 10. The progression of the geographic latitude (upper) and magnetic local time (lower) of the FAST apogee as a function of time since launch.

rest of the FAST mission, non-campaign telemetry contacts are 10–11 per day in the Northern Hemisphere and 4-6 contacts in the Southern Hemisphere.

For the first three years of operation, the FAST mission took advantage of the efficient multi-mission approach of the SMEX program at the Goddard Space Flight Center in which one flight operations team (FOT) is responsible for acquiring data from all of the SMEX missions. A multi-mission Payload Operations Control Center (POCC) at GSFC supports all SMEX missions (used by the SAMPEX, FAST, SWAS, WIRE, and TRACE satellites). The FAST Science Operations Center (SOC) is located at the University of California at Berkeley.

FAST relied initially on six geographically dispersed Ground Network (GN) or Deep Space Network (DSN) stations and the NASA Communications (NASCOM) network, as shown in Figure 11. These stations provide relatively transparent telemetry and command connections between the spacecraft and the Ground Data System (GDS) elements. The Poker Flat station in Alaska consists of the Transportable Operations and Tracking Station (TOTS), developed and operated by the Wallops Flight Facility (WFF) to collect real-time telemetry while the spacecraft is passing through the northern auroral region. A second TOTS, located at McMurdo, Antarctica, also acquires FAST data, yet less frequently. During the first winter campaign, northern hemisphere data were collected by the European Space Agency using a tracking station in Kiruna, Sweden in conjunction with a NASA-provided Ground Support Equipment (GSE) for FAST-specific telemetry processing. After the first

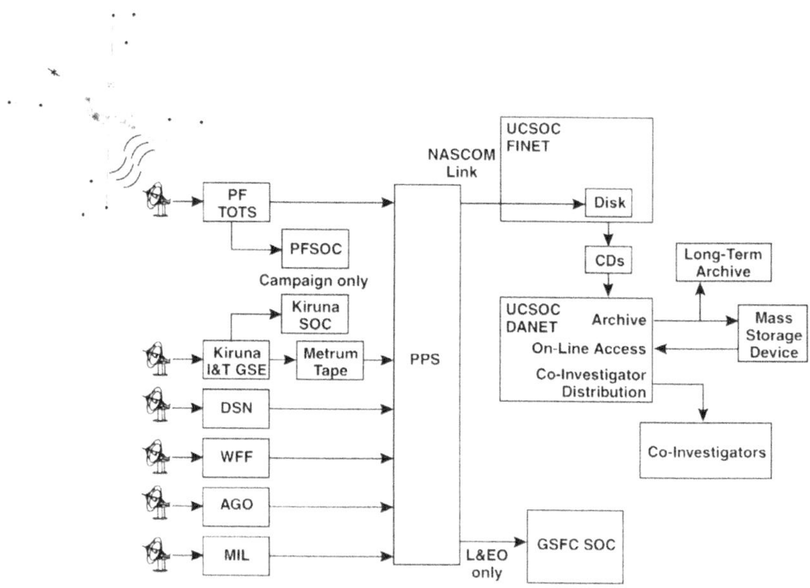

FAST Science Telemetry Data Flow

Figure 11. A diagram of the FAST operations and science telemetry flow for the first year of operations.

three years, FAST flight operations was transferred to the University of California at Berkeley where it is carried out in conjunction with the FAST SOC and with the (planned) HESSI flight operations.

FAST science downlinks can be performed at rates of 900 kbps, 1.5 Mbps, or 2.25 Mbps. In general, downlinks are carried out at the highest possible rates for a given contact. This mission scenario yields an average data volume ranging from 4–5 Gbytes per day (campaign) to 1–3 Gbytes per day (non-campaign).

FAST telemetry protocol follows design recommendations of the Consultative Committee for Space Data Systems (CCSDS) to provide a standard approach for implementing spacecraft telemetry and command interfaces. The CCSDS recommendations utilize layered packetization concepts (virtual channels, application-specific telemetry and command data packets), which simplifies the distribution of telemetry between the MUE, the IDPU, ground operational segments, the SOC, and the National Space Science Data Center (NSSDC).

An important aspect of the FAST mission is how it acquires its data and conducts its operations. The central ingredient to such decisions is the Science Operations Center at the Space Sciences Laboratory at the University of California, which is described in more detail by McFadden *et al.* (this issue). The FAST data flow follows a straightforward path from the satellite to the user, which is briefly outlined below.

The FAST satellite is fully capable of carrying out autonomous operations, storing data during one part of the orbit and telemetering it later during scheduled ground contacts. Furthermore, downloading the telemetry to unmanned ground stations ('blind dumps') is a routine feature of the FAST TOTS operations. All instrument commands are written at the University of California Science Operations Center (UCSOC) and ground, science, and housekeeping data are transferred to GSFC for minimal sorting by data type before being transferred to the SOC for analysis and archiving. Both the SOC and GSFC provide spacecraft and instrument health and safety monitoring and trending. (Note: This function is now carried out entirely at UCB after the first three years of operations).

All science operations are under the direction of the Principal Investigator, Dr. Charles W. Carlson. For the first three years, mission operations, planning and scheduling was the responsibility of the FAST Flight Operations Team (FOT) at Goddard. This FOT operated primarily out of the SMEX Mission Operations Room (MOR) at GSFC, using the CMS, POCC, and PPS systems previously described. These responsibilities shifted to UCB after the first three years.

The MUE flight software on-board the FAST spacecraft can receive and immediately act on commands (real-time command processing) or store them on-board for later execution (stored command processing). Stored commands can be activated based on either an Absolute Time Sequence (ATS) or a Relative Time Sequence (RTS). The latter are used primarily for event-driven autonomous spacecraft safing.

During all mission phases, each ground contact may begin with a command session, though nominally only one command session will be planned for a given day. The length of each contact will depend on the visibility of the ground station, but in no case will it exceed 30 min (limited by spacecraft power constraints). The amount of telemetry data received during each contact will depend on the length of the contact and the downlink data rate. Data on planned contacts is distributed to all involved parties via the Remote User System Terminal (RUST) system.

FAST science telemetry flow is also shown in Figure 11. The UCSOC archives all Level 0 data received from the PPS, provides on line access and software to perform data processing, and is the source for data sets to be archived by the NSSDC. Additionally, during campaign mission phases in the first two years of FAST operations, science telemetry flowed to various other temporary SOCs including one at the University of Alaska, Poker Flat Science Center and one at the Swedish Space Physics Institute (IRF) in Kiruna.

Initial instrument data processing for the FAST mission is accomplished with the Packet Processing System (PPS). The PPS is a highly automated system which performs all level zero processing and initial data distribution functions for the FAST mission. After the data are processed through the PPS, they are sent electronically to the SOC, where the data undergo further processing and are recorded on CDROMs.

The FAST data products consist of survey (or key parameter) data and high resolution level 0 data. The survey data are immediately available to the community via the World Wide Web (http://plasma2.ssl.berkeley.edu/fast/welcome2.html). They have a sampling resolution corresponding to once per spin period (5 s). The survey data are available from this website as GIF plots or as single orbit CDF (Common Data Format) files for plotting and use in other analysis programs.

The high resolution FAST data are analyzed using a program developed at the UC Space Sciences Laboratory called the Science Display Tool (SDT). Used by the entire FAST science team, the SDT program enables large quantities of data to be accessed very efficiently such that all data from all instruments may be displayed on the same screen in easily understood scientific units. The SDT program also interfaces to the Interactive Data Language (IDL) program which is used for further analysis of FAST data. To date, hundreds of IDL programs and modules have been written by the FAST Science Working Team (SWT) and are shared among all FAST data users. High resolution CDF files have recently become available on the FAST web site in addition to the survey data.

A key feature of the FAST observing strategy involves 'campaigns'. Campaign periods are the times during which data acquisition from the FAST spacecraft and concurrent measurements with other spacecraft, airplane, and/or ground based observations are emphasized. During these intense data observing periods, FAST high-resolution data are telemetered to the ground on every orbit. The first campaign lasted for 10 weeks beginning in January, 1997, when the FAST apogee was over the northern auroral zone. This time period represented a subset of the International Auroral Study (IAS) as part of the international Solar Terrestrial Physics Program (STEP), and was a period when auroral scientists concentrated on gathering simultaneous measurements from many different instruments around the world. Several additional observational campaigns have either been carried out or are planned that utilize radars, imagers, and magnetometers on the ground as well as other satellite platforms in space.

6. Summary

The FAST Small Explorer represents a dramatic advance over previous auroral satellites with regard to several important aspects of its mission, including its innovative scientific instruments, spacecraft design, flight software and operations, and data distribution. Taking advantage of the single P.I. approach inherent to NASA's Small Explorer Program, the success of the FAST mission rests upon a highly integrated spacecraft design and a highly centralized data management processing philosophy.

FAST's innovative design also includes sophisticated scientific instrumentation that were designed specifically for this satellite mission in order to gather extremely high time and spatial resolution fields and particle data. As the main scientific events are present only when the satellite passes through the Earth's auroral zones, the data are acquired by means of a unique on-board flight computer which controls all of the instruments and triggers burst data collection when the desired physical phenomena are encountered. In this manner, FAST provides the scientific community an unprecedented view of the auroral microphysics through such 'smart' data acquisition which occurs at exceedingly high rates. FAST is able to acquire such high time resolution data not simply because of the fast response of its instruments, but also because the electrostatic analyzers measure particles with continuous, uninterrupted 360° pitch-angle coverage that is independent of spin.

The highly successful FAST mission has far surpassed its minimum success criteria and is well on its way to providing the comprehensive statistical database necessary to significantly advance our understanding of auroral particle acceleration. Furthermore, by understanding the small scale 'microphysics' processes, a much clearer picture of the larger scale 'macrophysics' subsequently comes into focus, particularly in regards to the global system of auroral energy flow and large scale currents. In this manner, the FAST satellite is advancing our understanding of the Earth's aurora and related physical processes in a manner that far exceeds its original scientific objectives.

Acknowledgements

We acknowledge the support of Dr David Gilman of the Small Explorer Program Office at NASA Headquarters and of Drs E. Whipple, R. Carovillano, and L. Zanetti who served as Program Scientists for FAST during their respective tenures with the Space Physics Division of the Office of Space Science at NASA Headquarters. We are indebted to Mr Orlando Figueroa for the leadership he provided to the SMEX program during its initial phase at the Goddard Space Flight Center. We also acknowledge and thank Mr Tim Gehringer, who served for many years as the FAST Mission Manager and helped usher FAST through its early conception and design phases. Finally, we are indebted to the many engineers, technicians, programmers, and support staff both at Goddard and at the team member institutions, whose expertise, dedication, and hard work contributed immeasurably to the success of the FAST mission.

Work on FAST at the University of California, Berkeley, was supported through the NASA Small Explorer Project under Contracts NAS5-31283 and NAG5-3596.

References

Baker, D. N., Chin, G., and Pfaff, R. F.: 1991, 'NASA's Small Explorer Program', *Physics Today*, **44**, 44.

Carlson, C. W., Pfaff, R. F., and Watzin, J. G.: 1998, 'The Fast Auroral SnapshoT (FAST) mission', *Geophys. Res. Lett.* **25**, 2013.

Carlson, C. W., McFadden, J. P., Turin, P., Curtis, D. W., and Magoncelli, T.: 2001, 'The Electron and Ion Plasma Experiment for FAST', *Space Sci. Rev.*, (this issue).

Elphic, R. C., Means, J. D., Snare, R. C.,. Strangeway, R. J., Kepko, L., and Ergun, R. E.: 2001, 'Magnetic Field Instruments for the FAST Auroral Snapshot Explorer', *Space Science Reviews*, (this issue).

Ergun, R. E., Carlson, C. W., Mozer, F. S., Delory, G. T., Temerin, M., McFadden, J. P., Pankow, D., Abiad, R., Harvey, P., Wilkes, R., Primbsch, H., Elphic, R., Strangeway, R., Pfaff, R., and Cattell, C. A.: 2001, 'The FAST Satellite Fields Instrument', *Space Sci. Rev.*, (this issue).

Harvey, P. R., Curtis, D. W., Heetderks, H. D., Pankow, D., Rauch-Leiba, J. M., Wittenbrock, S. K., and McFadden, J. P.: 2001, 'The FAST Spacecraft Instrument Data Processing Unit', *Space Sci. Rev.*, (this issue).

Klumpar, D. M., Möbius, E., Kistler, L. M., Popecki, M., Hertzberg, E., Crocker, K., Granoff, M., Tang, Li, Carlson, C. W., McFadden, J., Klecker, B., Eberl, F., Künneth, E., Kästle, H., Ertl, M., Peterson, W. K., Shelley, E. G., and Hovestadt, D.: 2001, 'The Time-of-Flight Energy, Angle, Mass Spectrograph (TEAMS) Experiment for FAST', *Space Sci. Rev.*, (this issue).

Kruer, M. and Lyons, J.: 1994, 'The FAST Solar Array: Challenging Requirements, Novel Design', *Proceedings of the First World Conference on Photovoltaic Energy Conversion*, IEEE.

McFadden, J. P, Carlson, C. W., and Ergun, R. E.: 1999, 'Microstructure of the Auroral Acceleration Region as Observed by FAST', *J. of Geophys. Res.* **104**, 14453.

McFadden, J. P., Egun, R. E., Carlson, C. W., Herrick, W., Loran, J., Vernetti, J., Teitler, W., Bromund, K., and Quinn, T.: 1999, 'Science Operations and Data Handling for the FAST Satellite', *Space Sci. Rev.*, (this issue).

Mende, S. B., Heetderks, H., Frey, H. V., Lampton, M., Geller, S. P., Habraken, S., Renotte, E., Jamar, C., Rochus, P., Spann, J., Fuselier, S. A., Gerard, J.-C., Gladstone, G. R., Murphree, S., and Cogger, L.: 2000, 'Far Ultraviolet Imaging from the IMAGE spacecraft: 1. System Design', *Space Sci. Rev.*, **91**, 243.

Pankow, D., Besuner, R., Ullrich, R., and Wilkes, R.: 2001, 'Deployment Mechanisms on the FAST Satellite: Magnetometer, Radial Wire, & Axial Booms', *Space Sci. Rev.*, (this issue).

Parrish, Keith: 1999, 'On-Orbit Thermal Performance and Model Correlation of the Fast Auroral Snapshot Explorer', *Proceedings International Conference on Environmental Systems*, AES.

Schnurr, R., Everett, D., Gruner, T., Quinn, T., Chiville, M.: 1995, 'FAST MUE Power System Electronics: An Approach to a Magnetically Clean and Modular Design', Supplemental Proceedings of the 9th Annual AIA/USU Conference on Small Satellites, Logan, Utah, September 18–21.

THE ELECTRON AND ION PLASMA EXPERIMENT FOR FAST

C. W. CARLSON, J. P. McFADDEN, P. TURIN and D. W. CURTIS
Space Sciences Laboratory, University of California, Berkeley, CA 94720, U.S.A.

A. MAGONCELLI
Lockheed Martin, Advanced Technology Center, Palo Alto, CA 94304, U.S.A

Abstract. The ion and electron plasma experiment on the Fast Auroral Snapshot satellite (FAST) is designed to measure pitch-angle distributions of suprathermal auroral electrons and ions with high sensitivity, wide dynamic range, good energy and angular resolution, and exceptional time resolution. These measurements support the primary scientific goal of the FAST mission to understand the physical processes responsible for auroral particle acceleration and heating, and associated wave-particle interactions. The instrument includes a complement of 8 pairs of 'Top Hat' electrostatic analyzer heads with microchannel plate (MCP) electron multipliers and discrete anodes to provide angle resolved measurements. The analyzers are packaged in four instrument stacks, each containing four analyzers. These four stacks are equally spaced around the spacecraft spin plane. Analyzers mounted on opposite sides of the spacecraft operate in pairs such that their individual 180° fields of view combine to give an unobstructed 360° field of view in the spin plane. The earth's magnetic field is within a few degrees of the spin plane during most auroral crossings, so the time resolution for pitch-angle distribution measurements is independent of the spacecraft spin period. Two analyzer pairs serve as electron and ion spectrometers that obtain distributions of 48 energies at 32 angles every 78 ms. Their standard energy ranges are 4 eV to 32 keV for electrons and 3 eV to 24 keV for ions. These sensors also have deflection plates that can track the magnetic field direction within 10° of the spin plane to resolve narrow, magnetic field-aligned beams of electrons and ions. The remaining six analyzer pairs collectively function as an electron spectrograph, resolving distributions with 16 contiguous pitch-angle bins and a selectable trade-off of energy and time resolution. Two examples of possible operating modes are a maximum time resolution mode with 16 angles and 6 energies every 1.63 ms, or a maximum energy resolution mode with 16 angles and 48 energies every 13 ms. The instrument electronics include mcp pulse amplifiers and counters, high voltage supplies, command/data interface circuits, and diagnostic test circuits. All data formatting, commanding, timing and operational control of the plasma analyzer instrument are managed by a central instrument data processing unit (IDPU), which controls all of the FAST science instruments. The IDPU creates slower data modes by averaging the high rate measurements collected on the spacecraft. A flexible combination of burst mode data and slower 'survey' data are defined by IDPU software tables that can be revised by command uploads. Initial flight results demonstrate successful achievement of all measurement objectives.

1. Introduction

The primary scientific objectives of the FAST mission are to understand the microphysics of the aurora that lead to acceleration of electron and ion beams, heating of ion conics, and generation of a multitude of wave/particle interactions (Temerin

Space Science Reviews **98:** 33–66, 2001.

et al., 1990; Carlson *et al.*, 1998a). FAST instruments were specifically designed to obtain accurate, high resolution particle and field measurements on extremely short temporal and spatial scales. They are motivated by earlier instruments flown on sounding rockets that first resolved intense field-aligned electron beams and the rapidly modulated electron oscillations that cause flickering aurora (McFadden *et al.*, 1986, 1987). Recent measurements by the TESP instrument on Freja (Boehm *et al.*, 1994) also demonstrate the value of fast time resolution electron measurements at the lower edge of the acceleration region. FAST uses an integrated package of particle and fields instrumentation to make measurements at higher altitudes, within the heart of the auroral acceleration region, up to an apogee altitude of 4200 km. These measurements are revealing the origin of phenomena that were discovered on earlier spacecraft but not fully resolved or understood, such as suprathermal bursts and counterstreaming electron beams (Sharp *et al.*, 1980), ion conics (see, e.g., a review by Klumpar, 1986), auroral kilometric radiation (Gurnett *et al.*, 1974), and electrostatic shocks (Mozer *et al.*, 1977). Understanding wave-particle interactions quantitatively requires not only high time resolution, but also sufficient accuracy to compute wave growth and damping.

The FAST satellite is designed to acquire 'snapshots' or bursts of high speed data that cannot be transmitted continuously, as well as continuous 'survey data' covering the entire auroral pass. The instruments are designed to function as a single coordinated experiment to acquire complementary particle and fields measurements. A variety of data modes combine selected measurements from all of the instruments on several time scales. A typical auroral crossing includes 'slow survey' data where particle precipitation is low, 'fast survey' data when significant electron energy flux is detected, and 'burst' data with full time resolution when various criteria of high activity are found. The decision for switching modes is triggered by automatic on-board evaluation of the measurements. Additional discussion of the FAST data processing system and the fields data triggers are described by Harvey *et al.* (this issue) and Ergun *et al.* (this issue).

The FAST particle instruments include the electron and ion plasma instrument described here, which emphasizes rapid measurements of electron and ion pitch-angle distributions, and the TEAMS instrument (Klumpar *et al.* (this issue)), which provides complementary ion composition measurements of the full three dimensional distribution function of dominant auroral ions with a time resolution of 2.5 seconds. The following instrument description is presented in two parts. The overview section provides a general discussion of the instrument operation that is useful for interpreting the data products. The more technical information is presented in the later sections.

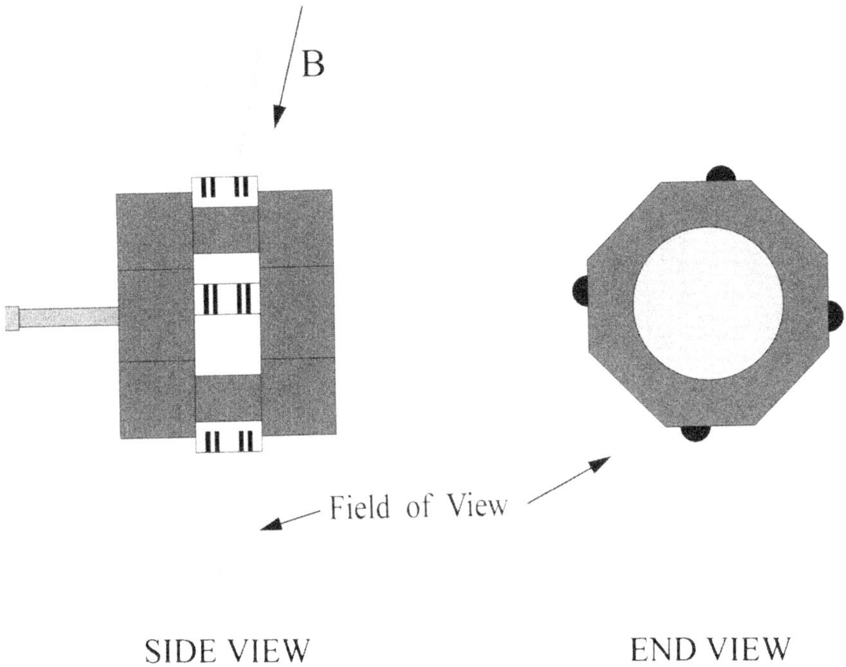

Figure 1. Two views of the FAST satellite illustrating the location of the four ESA stacks and their fields of view. Each stack contains four ESA sensors that are paired with a complementary stack on the opposite side of the deck to provide 360° radial fields of view. Deflectors at the entrance to the EESA and IESA pairs allow their fields of view to be aligned with the magnetic field as shown in the side view. The side view only illustrates four sensor fields of view for simplicity.

2. Instrumentation

2.1. Instrument Overview

The design challenge for the fast plasma instrument is to obtain continuous high resolution measurements of the electron and ion pitch-angle distributions with sub-second time resolution. Our design takes advantage of a property of the high inclination orbit to obtain continuous viewing of all pitch-angles. The FAST orbit has an 83° inclination, so the earth's dipole magnetic field forms a small angle to the orbital plane. The magnetic orientation of the FAST spacecraft axis is optimized by tilting the spacecraft spin axis out of the orbit normal by about 3° toward alignment with the earth's axis. With this orientation, the earth's magnetic field remains within 10° of the FAST spin plane throughout most high latitude auroral passes, so continuous pitch angle measurements can be accomplished by aligning detector fields of view with the spacecraft spin plane.

Figure 1 illustrates examples of representative fields of view of the FAST plasma instrument, which includes sixteen Top Hat electrostatic analyzers arranged in four 'stacks' of four analyzers each, mounted symmetrically around the central instrument deck. Each analyzer has a fan-shaped 180° field of view in the spacecraft spin

TABLE I

Fast particle detectors

Measurement	Energy range	$\Delta E/E$	Sampling resolution	Field of view	Angular resolution	Sample array	Geometric factor (cm^2 sr E/E)
Ion spectrometer	5 eV–24 keV	0.20	78 ms	360° × 12°*	11.2° × 12°	48E × 32α	0.009
Electron spectrometer	6 eV–30 keV	0.15	78 ms	360° × 10°*	11.2° × 10°	48E × 32α	0.005
Electron Spectrograph	6 eV–30 keV	0.15	1.6 ms	360° × 10°	22.5° × 10°	6E × 16α**	6 × 0.01

*Field of view can deflected ±10 deg from the center angle.
**A larger energy sample array can be achieved with poorer time resolution.

plane, and analyzers located on opposite sides of the spacecraft operate as pairs to obtain a full 360° view. The main solar panels are body-mounted on octagonal structures located above and below the instrument deck. Additional small solar panels fill areas between instruments on the instrument deck.

For simplicity, the magnetometer and electric field booms are omitted from this figure. These booms mount to the instrument deck and are located to cause minimum interference with the particle detectors. The electric field booms in the spin plane are thin wires that are located between the entrance apertures of the analyzers. The exterior conductor of the boom is at spacecraft potential and the wire obscures a negligible portion of the field of view. Likewise, the magnetometer booms are located to obscure a minimum solid-angle within the plasma instrument field of view. Further description and diagrams of the booms and their configuration on the spacecraft are presented in companion papers (Pfaff *et al.* (this issue); Pankow *et al.* (this issue)).

The FAST plasma analyzers use a common design that is configured for three specific applications to optimize either high energy and angle resolution or high time resolution. The basic analyzer parameters for these three options are summarized in Table I. Two of the analyzer pairs operate as swept energy spectrometers to measure electron (EESA) and ion (IESA) distributions, emphasizing energy and pitch-angle resolution. Each of the detector stacks includes one of these spectrometer analyzers, which have 16 azimuth angle sectors and typically sample 48 energy bins spanning energies from 4 eV to 32 keV for electrons and 3 eV to 24 keV for ions. Spacing of energy samples is designed for approximately constant $\Delta E/E$ spectral resolution. The nominal energy sweep rate of 12.8 sweeps s^{-1} is phase-locked to produce exactly 64 sweeps during the 5-s spin period of the spacecraft. The standard accumulation time at each energy step is approximately 1.63 ms. Other selectable sweep options include a double-speed mode with 24 energy steps, and a half-speed, high resolution sweep with 96 steps. Each sweep cycle requires one retrace step so the number of resolved energy levels is one fewer than the number of steps in the sweep.

The spectrometers include deflection plates that independently steer the field of view of each analyzer up to 10° above or below the spin plane to track the magnetic field direction, creating continuous coverage of all pitch angles. As Figure 1 illustrates, the deflections of the upward and downward viewing analyzers are in opposite directions to allow simultaneous viewing in both directions along the magnetic field. This capability is essential for investigating magnetic field-aligned electron and ion beams, which may travel in either direction along the magnetic field. The central Instrument Data Processor Unit (IDPU), (see Harvey *et al.*, this issue), uses magnetometer data to automatically compute deflection angles and command the analyzer deflector plate voltages.

The geometric factors for the EESA and IESA sensors listed in Table I are consistent with the energy sweep times and counting rate capabilities of the detector electronics. The electron differential energy fluxes of auroral electrons typically range between 10^7 and 10^{10} eV/cm^2-s-sr-eV. Each anode in the EESA has a geometric factor of 1.6×10^{-4} cm^2-sr-eV/eV, so typical count-rates range between 10^3 and 10^6 per second per anode. The pulse amplifiers have a deadtime of 140 ns, so these rates are easily accommodated, and the MCP gain and strip current are designed for 5 MHz rates. The standard accumulation time of 1.6 ms thus obtains samples ranging between 1 and 10^3 counts. The accumulator circuits have 14 bits, for a capacity of 1.6×10^4 counts. Lower fluxes are easily identified by averaging multiple sweeps. The IESA sensor has a somewhat larger geometric factor to match the lower expected ion fluxes.

The EESA/IESA sensors provide a good combination of energy, angle, and time resolution that form the baseline plasma particle measurements. They are typically operated in a standard mode that covers the entire energy range at full time resolution. The total volume of data are compressed for summary data by averaging these measurements in the IDPU.

The remaining six pairs of analyzers are electron sensors that collectively function as a spectrograph (SESA), using a variety of programmable energy scan cycles to accomplish the highest time resolution measurements. Each of the SESA analyzers has 16 pitch angle sectors and can operate over the same 48 step energy range as the EESA spectrometer and uses the standard 1.63 ms accumulation cycle. This parallel detection design uses a total of 96 separate angle sector anodes with individual amplifier/accumulator strings. The large geometric factor (0.06 cm^2-sr-eV/eV) of the SESA instrument is an order of magnitude larger than the EESA, allowing it to make electron distribution measurements with unprecedented time resolution and good counting statistics. The geometric factor for each SESA anode is four times larger than for the EESA. Thus the count-rates for intense beams with fluxes of 10^{10} eV/cm^2-s-sr-eV will be 4×10^6 per anode, which is consistent with the design limit of the amplifiers, accumulators, and MCP saturation level.

TABLE II
Weight and power

	Per Stack	Total
Mass	3.85 kg	15.4 kg
Power:		
LV	1.8 W	7.2 W
HV	1.2 W	4.8 W
Total	3.0 W	12 W

A variety of hardware programmed operating modes are used with the SESA. They can operate in a full energy spectrometer mode, similar to the EESA, with each analyzer sweeping over the complete energy range, but with a phase shift between each of the six pairs. This mode generates full energy spectra every 13 ms, six times faster than EESA, but with half the angle resolution. Faster modes are accomplished by dividing the energy range among the six analyzer pairs and reducing the number of energy steps for each analyzer. For example, a 24 step energy spectra (4 steps per analyzer) can be collected every 6.5 ms, a 12 step spectrum requires 3.3 ms, or each SESA pair can remain at fixed energies, yielding a six point spectrum every 1.63 ms. The energy range for these cycle modes can be selected to span the entire energy range, or can be concentrated in narrower ranges. One application using a narrow energy range is the 'peak tracking mode', which uses the EESA data to identify a peak in the energy distribution and then slaves the SESA to make rapid measurements in a limited energy range surrounding the peak. Table I summarizes the possible combinations of energy and time resolution for the EESA and SESA.

Operating modes for the ESA sensors are programmed in hardware. The raw measurements produced by the ESA detectors represent the highest speed burst data that are transmitted to the ground. Lower rate survey data products are generated in the IDPU by time-averaging particle distributions according to selected software modes. The ESA hardware modes are typically fixed throughout a single auroral zone pass, whereas the IDPU may have several levels of data compression that are adaptively selected according to onboard data evaluation algorithms. For example, the fast survey, slow survey, and burst modes are determined by evaluations of measured data performed by the IDPU. The IDPU always generates the plasma survey data with at least two distributions per spin that are available at one spin resolution in CDF (Common Data Format) files available from the NSSDC (National Space Science Data Center) or as summary data plots from the FAST web site.

2.2. Design considerations

The FAST mission imposed severe constraints on the spacecraft and instrument designs. The science requirements demanded very capable instruments generating exceptionally high data rates that would fit within the cost and launch capability of a Small Explorer mission. The instrument designs had to minimize weight, power, and volume, and tolerate a high radiation environment to attain the desired orbit. Rather than a 'box on deck' design, FAST instruments followed a more integrated design where instrument boxes and solar panels form a complete radiation and thermal close-out of the instrument deck. With this design, interior components receive a minimum radiation shielding equivalent of 4 mm of aluminum. Since the plasma instrument must be exposed to the outside, its design ensured that all electronic components were located behind passive structure or explicit shielding. The electric field boom mechanisms are also integrated with the analyzer stacks to increase shielding, conserve weight, and provide a single interface to the spacecraft. The weight and power requirements for individual stacks and for the total instrument are listed in Table II.

Because the plasma instrument uses a large number of analyzers, their design criteria emphasized ease of fabrication and testing. The detectors are composed of modular components that can be assembled and tested separately and are easily replaced if necessary. The design also provides contamination control features including sealed apertures and a dry nitrogen purge system.

2.3. Detector design

The main elements of the plasma detector design are the particle optics design, the mechanical design of the electrostatic analyzers, the electrical and mechanical design of the MCP anode boards, and the contamination control system.

2.3.1. *Analyzer Optics*

The electrostatic analyzers used on FAST are spherical section top-hat analyzers (Carlson *et al.*, 1983; Carlson and McFadden, 1998) with a uniform 180° field of view. As shown in Figure 2, each analyzer consists of an inner hemisphere mounted on an insulating plastic ring, and a concentric outer hemisphere with a circular entrance hole at the top. A third concentric top-hat electrode is suspended above the entrance hole. An attracting voltage is applied to the inner hemisphere which allows charged particles with the corresponding energy per charge to pass through the analyzer and be detected at the microchannel plate (MCP) electron multiplier. An entrance aperture in the cylindrical analyzer housing collimates the particle beam and also restricts entry of solar UV radiation into the analyzer. Collimation apertures are also located near the entrance to the top-hat region, and the top-hat entrance itself defines the final collimation. Conical deflection plates are included behind the entrance collimator of the EESA/IESA analyzers to steer the detector field of view by ±10°. Typical particle trajectories in each analyzer are shown.

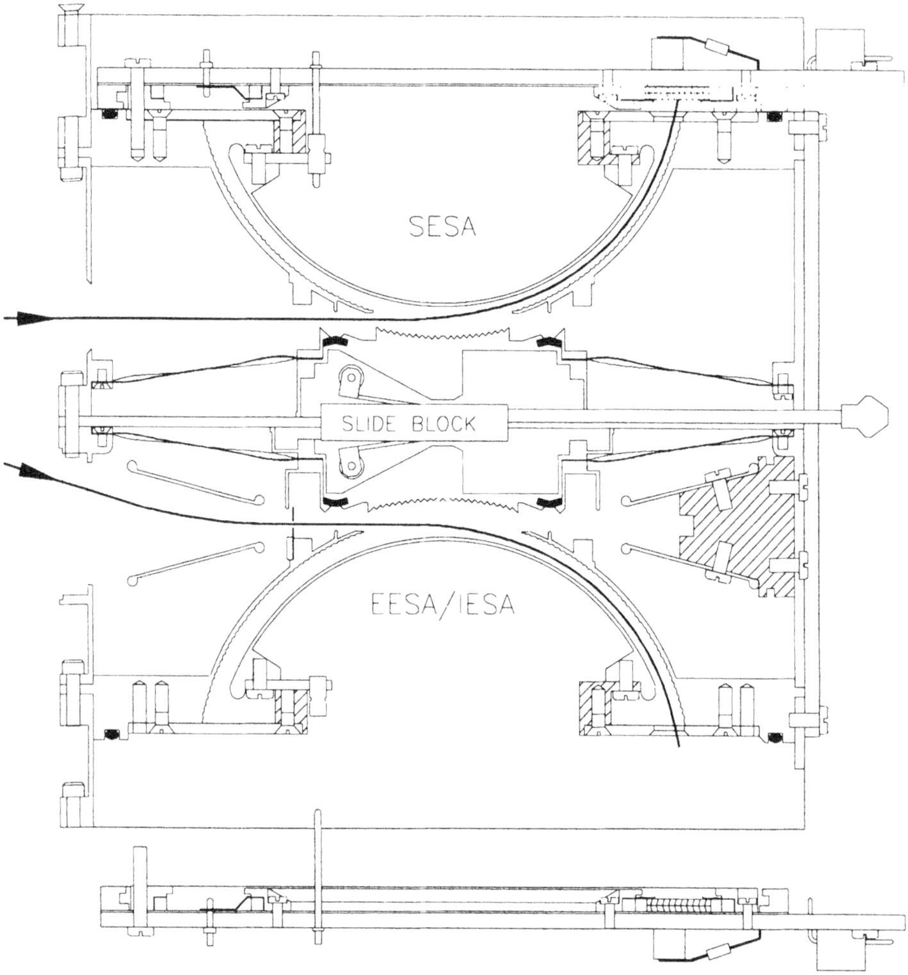

Figure 2. A cut-away drawing of an ESA pair module, illustrating the hemisphere mounting configuration, the top-hat closing mechanism, and the MCP anode mounting. Particle trajectories show the deflection through the analyzer, and also the operation of the entrance deflection plates that allow these sensors to continuously measure magnetic field-aligned fluxes.

Because negatively charged surfaces might generate accelerated photoelectrons, the deflection systems use only positive voltages to either attract electrons or repel ions.

The particle optics response for the FAST analyzers was thoroughly characterized by computer simulations (McFadden and Carlson, 1998) and with vacuum chamber calibrations. The simulated response, plotted in Figure 3(a), illustrates the typical angle/energy skew expected for a 90° deflection analyzer. Simulations are compared with measured data for integrated energy and angle responses in Figures 3(b) and 3(c). The energy response in Figure 3(b) was measured with an electron beam energy of 10 keV. The overlapping traces at the bottom of the

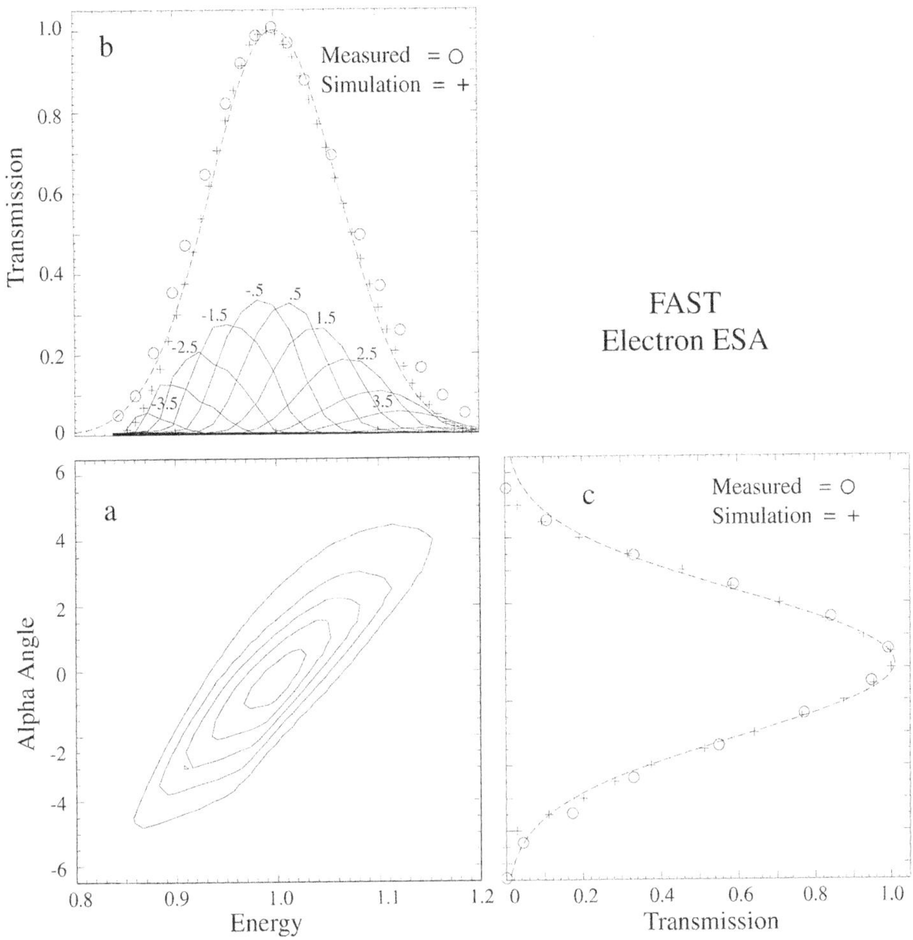

Figure 3. (a) The FAST EESA energy-angle response. The energy-angle contour plot is the simulated analyzer response. (b) The simulated and measured energy response averaged over angles. (c) The simulated and measured angle response averaged over incident energy.

plot are energy scans at each 1° increment of elevation angle α. The measured response integrated over α is plotted as open circles. The computer simulation data are shown with + symbols, which have also been fitted with a Gaussian function denoted with a dashed line. The measured response shows good agreement with the simulation, but is slightly wider on the high energy wing. Measurements using ion beams did not show this widening, so we attribute the wider electron response to small angle forward scattering of electrons hitting the outer hemisphere. The analyzer response parameters summarized in Table I are derived from the analyzer simulations.

2.3.2. *Analyzer Mechanical Design*

The most critical mechanical requirement for the analyzer design is the stable and accurate alignment of the three hemispherical electrode surfaces. Requirements from other considerations are ease of fabrication and a simple interface to other components making up the complete instrument. The inner and outer hemispheres and the top-hats are machined from aluminum. Figure 2 illustrates most of the mechanical components of the analyzers. The electrode surfaces of the top-hats and the outer hemispheres have a scalloped finish that reduces scattering of particles and UV photons that can reach the exit aperture and produce background events. The hemispheres and top-hat components were gold plated to ensure good surface electrical properties, and a gold-black coating was vacuum deposited on their electrode surfaces to further reduce particle and UV scattering. Other analyzer surfaces are painted with conductive paint to ensure good surface conductivity.

The mechanical alignment of the analyzer components is maintained by using a small number of simple parts in the optical system. The outer hemisphere is machined from a single piece of aluminum that also forms the main mechanical structure of the analyzer assembly. A flat circular exit aperture plate is fastened to the hemisphere base and extends under the inner hemisphere. The aperture plate has an annular exit aperture that extends approximately 180°. An etched BeCu grid covers the exit aperture to terminate the analyzer electric field at the aperture plane. An insulating ring made of PEEK connects the inner hemisphere to the aperture plate. PEEK is a very stable, high temperature plastic with low outgassing properties. The plastic ring has a series of vertical slots machined about half way through the vertical height of the ring to accommodate differential thermal expansion between it and the aluminum parts without affecting the axial alignment. The fabrication tolerances for the hemispheres, aperture plate, and plastic ring are tightly controlled, and the combination of flat and cylindrical alignment surfaces provide stable positioning in both the axial and radial directions.

The alignment of the top-hat is also determined by a simple mechanical path. The outer hemisphere base is fastened directly to the external cylindrical tube that contains the entrance aperture collimator. A single flat aluminum plate fastens to the top of the tube and forms the common interface between the two halves of the analyzer module. A spoked spring suspends and centers the top-hat assembly within the cylinder housing. This suspension spring pulls the top-hat upward against the center plate, which determines the vertical position of the top-hat. The suspension spring is part of the aperture closing mechanism that is described later. The lower analyzer in Figure 2 illustrates the deflection plate mounting for the I/EESA. Two blocks of PEEK plastic provide a cantilever mounting of the analyzer plate assembly to the cylinder body. A double mesh grid covers the entrance aperture at the front of the analyzer housing to define the electrical potential boundary conditions for the deflector optics and also prevent leakage of the potential into the external plasma.

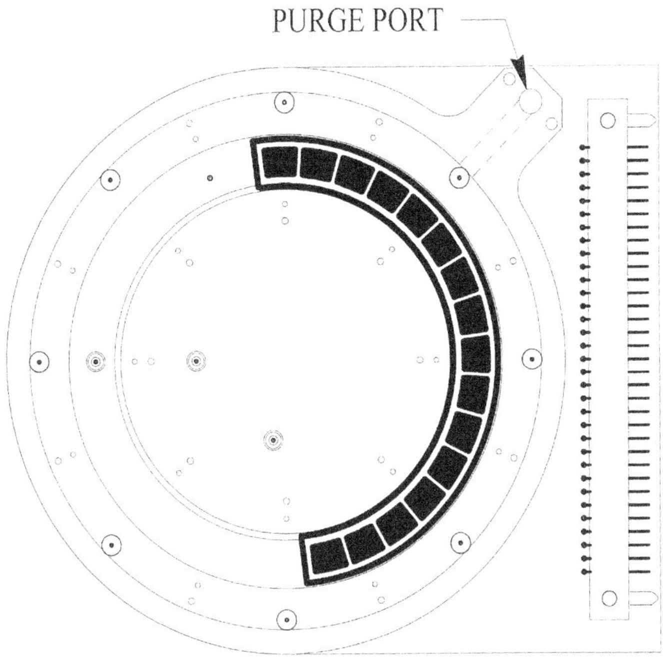

Figure 4. A mechanical drawing of the EESA/IESA anode board showing the configuration of the charge collection pads, the MCP electrical contact, and the purge port used for the nitrogen purge system.

The final components comprising the structure of the analyzer assembly are the top and bottom interface rings. These components have a 'D' shape when viewed along the analyzer axis, and they serve to join the cylindrical analyzer assemblies to the rectangular housing for the detector electronics. They are also the housing for the MCP anode board assembly, which is shown in two configurations in Figure 2. The lower anode board is detached from the analyzer to show how it appears as a separate subassembly. It is normally installed in the analyzer with its MCP aligned with the exit aperture. The anode board also has a pin that connects the sweep high voltage to a socket on the inner hemisphere. The top analyzer is shown with its anode board installed. A Viton 0-ring makes a seal between the anode board and the analyzer base for contamination control.

2.3.3. *MCP Anode Board Assembly*

The anode board assembly includes the MCP mounting hardware, the signal anode pads, and the electrical circuitry needed to route the signals and high voltages to connectors. It is designed and fabricated using standard multilayer printed circuit technology. The substrate is a low outgassing polyimide board material that can be baked out at high temperatures.

The anode patterns and the general mechanical layout of the 16 anode EESA board is shown in Figure 4. The anode pads with their signal routing traces are

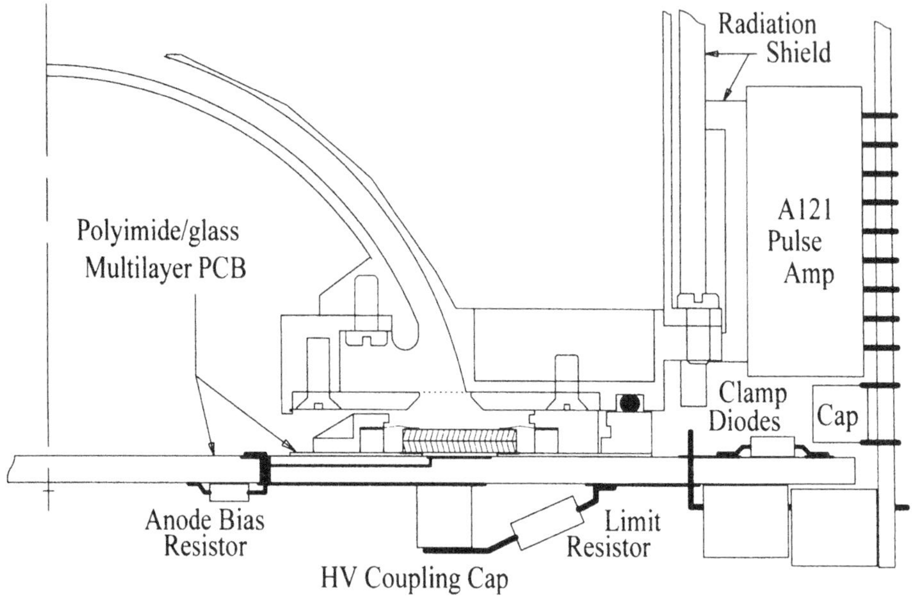

Figure 5. A cut-away view of the ESA analyzer showing the mechanical mounting of the MCPs and the electrical signal path from the anodes to the preamplifiers.

located on an internal layer of the pc card. A semi-circular window through the top layer of the pcb exposes the anode pads to the MCP, which mounts on the metal trace surrounding the anode window. The thickness of the top layer of the pcb lamination is 0.4 mm thick, which establishes the gap between the MCP and anode surface. The cut-away view in Figure 5 shows mechanical and electrical details of the anode board and its interfaces to the electrostatic analyzer and preamp electronics board. The anode pads are located at the bottom of the small gap below the MCPs. A buried trace connects each anode to a bias resistor that provides the dc path for anode current. The signal then passes through a HV coupling capacitor and a surge limit resistor to the signal connector that mates with the pulse amplifier board. The coupling capacitors are omitted for the ion instruments since the anodes are at low DC potential. Two diodes provide a bipolar voltage clamp on each signal trace to suppress any voltage spikes caused by sparking or corona. This figure also illustrates the close proximity of the pulse amplifiers to the anodes, their radiation 'spot shielding', and the mounting of the amplifier boards to the analyzer modules. High voltage connectors on each anode board plug directly into high voltage supplies. They are not included in Figure 5 but are shown in the exploded assembly drawing (Figure 6).

The two MCPs are mounted as a chevron pair with no inter-plate gap. The plates have 25 micron pores with a length/diameter ratio of 40:1. A rather high strip current of 5 μA cm^{-2} is chosen for high count-rate capability. Imaging applications often include a gap between the plates to increase the MCP gain, but

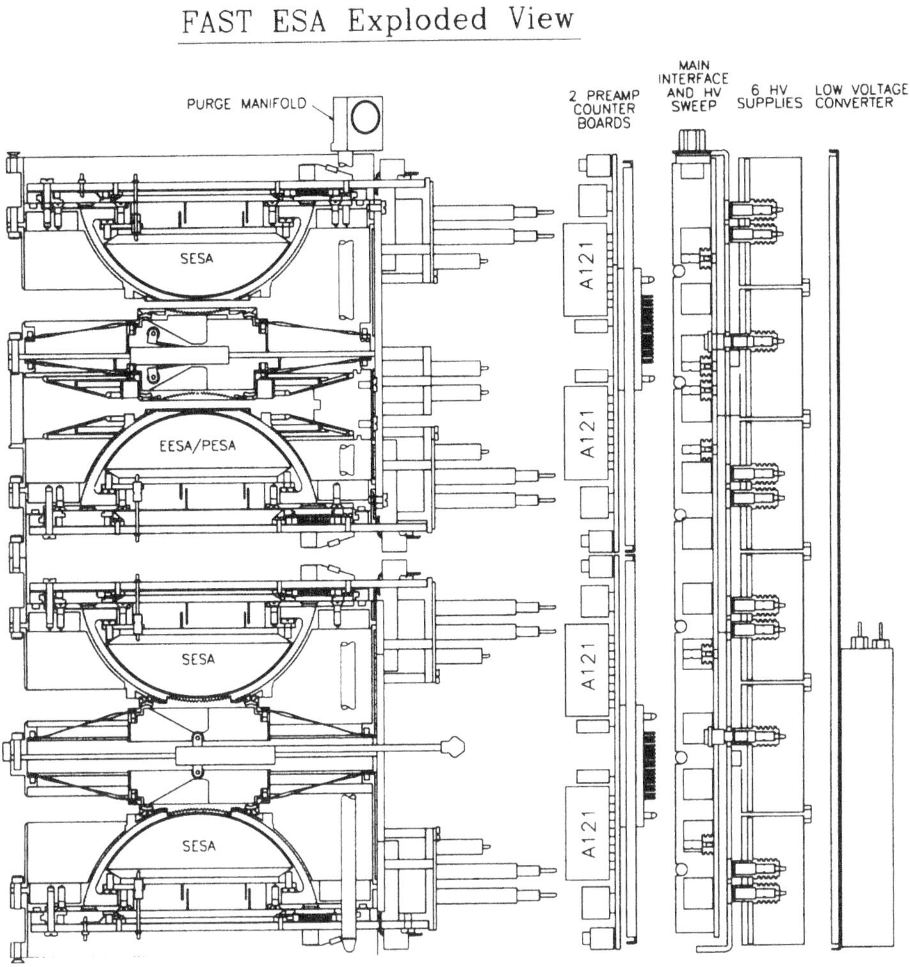

Figure 6. An exploded view of the FAST ESA stack assembly showing the stack of four analyzers, preamplifier boards, the interface and HV sweep control board, and the high and low voltage power converters. The modular design allows easy assembly for servicing and refurbishment.

that configuration also degrades the pulse-height distribution and increases gain variations between different positions on the plate. Image resolution for the FAST system is determined by individual anodes, and the design goal is maximum count rate capability. For these conditions excess MCP gain is a disadvantage, because it reduces the maximum count rate before onset of MCP saturation. The tightly clamped MCP configuration used for FAST optimizes uniform electron gain and narrow pulse height distributions with a modal gain of $1-2 \times 10^6$ electrons. The MCP and preamps are designed to handle maximum count rates of 5 Mhz per anode segment without significant drop in detection efficiency.

The details of the MCP mechanical mounting are shown in Figure 5. Circular plastic rings, screwed to the anode board on each side of the MCP annulus, hold rings of Be-Cu spring fingers that clamp the MCP to the board. This clamping holds the MCPs in tight contact and also provides a good thermal path from the plates to the anode board. Because the channelplates have a negative thermal coefficient of resistance, high current applications such as FAST must be designed carefully to avoid thermal runaway. The outer MCP mounting ring is also the surface that contacts the analyzer aperture plate when the anode board is installed. The screws attaching the anode board to the analyzer base pass through this mounting ring. A concentric polyimide stiffener ring is bonded to the anode board just outside the outer mounting ring. An o-ring seals the joint between the stiffener ring and the analyzer base, so the only opening to the volume surrounding the MCP is the top-hat aperture. There is also access to introduce purge gas through a radial hole drilled in the edge of the stiffener ring (see Figure 4). Details of the purge system are presented in the next section.

2.3.4. *Contamination Control*

Microchannel plates are susceptible to contamination by particulates, and are also damaged by absorption of water vapor. The best storage environment for the MCP is either in a vacuum or a purged 'dry box'. The FAST analyzer modules provide similar protection for the MCP during storage and launch by including a nitrogen purge system for the sensors and using the top-hat assembly to seal the interior volume of the analyzer.

Figure 2 shows the main elements of the top-hat closer mechanism. Suspension springs center the two top-hats and pull them against the central mounting plate to hold them open during normal operation. The closer mechanism has spring-loaded rollers mounted on a slide-block that push against the inclined ramps machined into the interior of the top-hats. Horizontal motion of the slide-block thus pushes the two top-hats outward until the annular gaskets in the top-hats make contact and seal against thin circular protrusions on the outer hemisphere surface. The spring force on the rollers allows the top-hats to vent gas at a moderate over-pressure. The curved top-hat electrode protrudes into the hole in the outer hemisphere when the the closer is engaged. The slide-block is attached to a spring-loaded actuator rod that is released by a melt wire to open the top-hat seal on orbit.

The MCPs are considered the most vulnerable component for contamination, so the purge system is designed to deliver purge gas directly to the interior of the analyzer via the anode board. The spacecraft includes plumbing to distribute pressurized N_2 gas to the particle instruments. The gas is supplied to each analyzer through a metal 'skewer' tube that passes though a small o-ring gland mounted on the corner of each anode board (see Figure 4). The skewer tube has small manifold holes aligned with each of these glands. The purge channel in each anode board contains a sintered filter and a small metering orifice that regulates the gas flow rate to each analyzer.

2.3.5. *Stack Assembly*

Each of the four instrument stack assemblies contains three SESAs and either an ion or electron spectrometer (IESA or EESA). The stacks are composed of two major sub-assemblies that can be assembled and tested individually, and the design of each sub-assembly is also modular. The exploded view of the instrument stack (Figure 6) illustrates the modules that make up the instrument stack. The analyzers are grouped in two dual analyzer modules with anode boards. This figure shows the anode board with their high voltage plugs that were missing from Figure 5. A counter/preamplifier board is mechanically mounted to each analyzer module and has connectors that mate to the two anode boards. A thin aluminum sheet is mounted to the top surface of the preamp board to shield it from electromagnetic noise on the main interface board. The sensor subassembly can be tested as a unit using external high voltage supplies and a simple digital test interface. Initial vacuum chamber testing of the analyzer assembly was done at this level.

The second major subassembly includes the high voltage and interface functions and contains seven circuit cards mounted on a common structural panel. One board includes the high voltage sweep generator, the main instrument interface circuitry, and the spacecraft interface connector. The other side of the mounting panel holds six high voltage converters with their interface mother-board. Two of these converters supply fixed high voltages to the sweep generator circuits. These connections are made by high voltage plugs and sockets mounted directly on the boards. The final component shown in the exploded view is the low voltage power converter mounted to a shield plate, which mounts above the high voltage supplies on metal posts. The electrical connections between the power converter and interface board are made with the only cable harness in the stack.

The stack subassemblies are mated by carefully sliding the two units together. The preamp boards have interface connectors that plug directly to mating connectors on the interface card. The anode plate high voltage plugs connect to sockets on either the interface board or the high voltage boards. The longer plugs pass through clearance holes in the interface board to reach the HV supplies.

The mechanical structure and enclosure of the stack is shared on one side with the electric field wire boom unit, which shares a vertical structure plate with the ESA stack. The back cover is a plate that includes support electronics for the wire boom unit. The second side is also a structural plate, and the top and bottom are made of thin aluminum sheets.

2.4. Electronics Design

The electronic systems for each of the four ESA stacks are nearly identical and are summarized in the block diagram in Figure 7. They consist of four anode boards, two amplifier/counter boards, one board combining the main system interface functions and the HV sweep generator circuits, six high voltage converters, and the LV converter. The timing and programmable functions on these boards

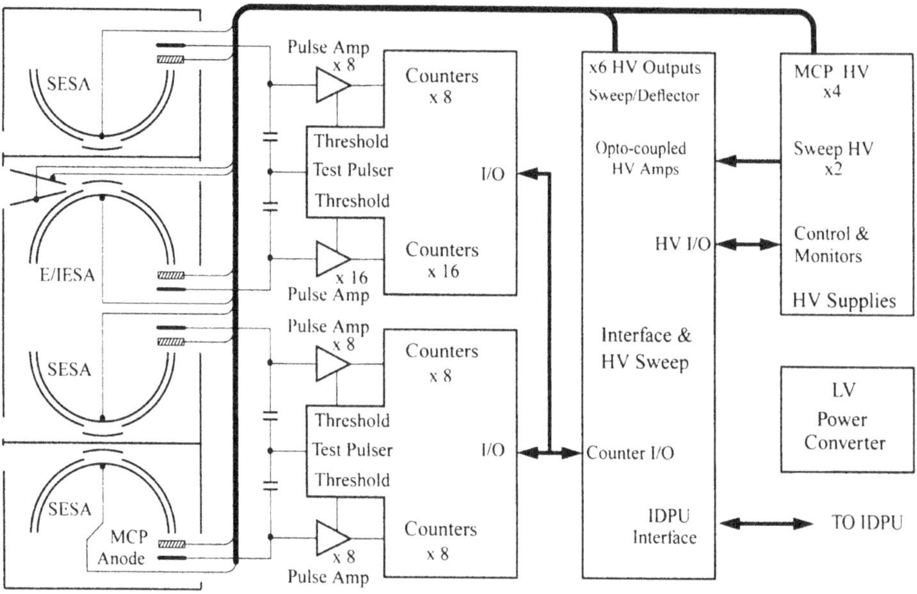

Figure 7. Block diagram of the electronics for a FAST ESA stack. The location of electrical component blocks are in the same relative positions as their mechanical counterparts in Figure 6.

are controlled by commands from the IDPU. All logic functions as well as certain analog functions are implemented in nine Actel A-1020 field programmable gate arrays (FPGA). These devices were an enabling technology for FAST. It is a single part type that is radiation resistant, consumes very little power, and replaces the functionality of approximately forty standard integrated circuits. The gate array can also be used for certain mixed digital/analog designs. All of the digital to analog converters in the FAST instruments are implemented by simply powering the FPGA with an accurate 5 Volt power source and driving standard R/2R resistor networks with the output logic gates.

The IDPU controls the operation of the ESA instrument with a serial I/O command interface. A spin-synchronous command gate recurring 12,288 pulses spin^{-1} serves as the master time base for the instrument measurement cycle. A 24-bit command data block containing 8 bits of address header and a 16-bit data word can be sent with each command gate pulse. The IDPU routinely issues 'broadcast' commands to all instruments every spin that contain spacecraft time, a 'spin sync' pulse once per spin, and 'magnetic phase' data that identify the phase angle between the spin sync and the location of the magnetic field plane. The commands for sweep and data format modes are also broadcast commands that go to all ESA stacks simultaneously. In addition to these critical timing and attitude data, the IDPU sends addressed commands to specific instruments. All configuration commands to the ESA stacks that control specific functions, such as the MCP voltage level for a specific sensor, are addressed commands.

The data readout from the ESAs to the IDPU is a fixed format that is also synchronized to the spin. Each ESA stack has its own hardware interface to the IDPU, and data transfers are made at a clock rate of 1.048 Mhz from all four stacks in parallel. Thus the data samples for a specific anode angle and energy from all four stacks arrive at the IDPU simultaneously and can be efficiently packed in a single data array. The IDPU reduces the 14 bit counter data to 8 bits via a pseudo square root compression.

Analog housekeeping signals are routed directly to the IDPU for digitization and formatting with other engineering data.

2.4.1. *MCP/Anode Board*

Some of the electrical functions of the anode board were already described in Section 2.3.3. The two main electrical design issues for the anodes are the high voltage power for the MCPs and the anode signal output coupling. The MCPs operate with a cross-plate potential of approximately 2 kV. The detailed high voltage configuration is different for the electron and ion sensors.

The ion MCPs operate with the input surface at high negative potential, which accelerates incident ions to improve their detection efficiency. The anodes operate at ground potential, and the output face of the MCP is biased at approximately 25 V negative to ground by a series resistor in its ground return. This slight acceleration voltage improves the focusing and collection of the output electron charge cloud by the anode. The electrical connection to the front and rear of all MCPs are bypassed with filter capacitors directly to the signal ground plane on the anode board.

The electron anodes operate with approximately 600 V of pre-acceleration voltage to ensure good detection efficiency for low energy electrons. A series resistor could generate this voltage, but it would dissipate approximately 25% of the MCP power. The FAST design conserves this power by tapping the pre-acceleraton voltage from the first stage of a five stage voltage multiplier in the HV supply. The output electron cloud is post-accelerated by about 25 V by an MCP series resistor. Because the electron anodes operate at high voltage, the output signal is coupled through monolithic ceramic capacitors that are surface mounted to the anode with silver epoxy.

Two anode versions are used. The board for the EESA and IESA analyzers uses 16 anode pads to cover the 180° annulus. The component loading is different for these two versions, but the same anode design is used. The SESA anode has 8 anodes, each with 22.5° angular width. The MCPs and their mounting hardware are identical on all anode types.

The anode electrical design resists crosstalk and noise pickup by the anode electrodes. An internal ground-plane shields the anode pads from each other and from other signals on the back surface of the board. Alternate pins on the signal connector joining the anode board with the preamplifier circuit board are also grounded. No crosstalk between anodes pads was found over the entire range of MCP pulse amplitudes. The design is also resistant to external noise pickup. A small amount of duplicate event counting is introduced at the boundary of anodes, where the electron cloud from the MCP spreads to anode pads on each side of the boundary. The gaps between the anode pads and the gap between the MCP and the anode surfaces are much smaller the anode width, so this double counting contributes only a few percent to the total rate.

2.4.2. *Amplifier/Counter Boards*

The MCP output signals are amplified, discriminated, and accumulated by circuitry on the counter boards. The pulse amplifiers are the AMPTEK model A121, with their dead-times adjusted to 140 ns. These amplifiers have an adjustable threshold that can be set by ground command. The threshold can be adjusted for optimum noise rejection, and is also used to measure pulse height distributions of the MCPs on orbit. A diagnostic function included on this board is a programmable tail pulser that is used for in-flight testing of the preamps.

Two versions of the counter board were required, since the two SESA analyzers are identical with 8 anodes each, while the second pair of analyzers has 16 anodes on the I/EESA analyzers. All digital functions are implemented in Actel A-1020 programmable gate arrays. Three configurations of the gate array are used on the counter boards. The basic functions performed by these circuits include the following:

(1) A Decoder for commands received from the IDPU.

(2) A 14 bit counter, latch and readout multiplexer for each of 8 anode channels. Also includes a programmable input mask to disable noisy anode channels.

(3) A Despun channel selector to send raw counts from any addressed pitch-angle channel to the IDPU (used for the wave/particle correlator).

(4) A Despun channel selector to send raw counts from the channel aligned with magnetic field direction (used for real-time energy flux computation).

(5) A D/A converter to set the amplitude for a test pulser; also a tail-pulse generator switch.

These functions are implemented in three specific FPGA circuit designs. The first circuit is the master device that includes the command decoder, the output data interface, an 8 channel counter circuit, and the data selector for pitch-angle selected counts. The second FPGA contains an 8 channel counter circuit, the data selector for magnetic field-aligned counts, and the test pulser composed of a D/A converter and switch. These two gate arrays are used on both versions of the counter board. A third FPGA design is used only for the I/EESA board that contains the extra 8 channel counter circuit.

The block diagram shows the test pulser capacitively coupled to the preamp in parallel with the anode input. The pulser signal line is simply routed perpendicular to the input traces on the preamp board, with the trace widths selected to provide a fraction of a picofarad of coupling. The threshold control voltages for the preamplifiers are generated by A/D converters on the main interface board and routed to the counter boards.

The A121 preamps are hybrid in-line packages that are mounted vertically on the circuit board as shown in Figure 5. An aluminum channel runs across the top of each row of amplifiers for radiation shielding and vibration support. It was attached to the amplifiers with staking material and anchored to the pbc with standoff posts.

The preamp board plugs directly onto connectors on both anode boards in the analyzer set, and is mechanically mounted to the structural plate on the analyzer. Each preamp board plugs directly to the main interface board with a board-mounted connector.

2.4.3. *Interface/HV Sweep Generator Board*

The interface and HV sweep functions are relatively independent functions that are combined on a single circuit card for the convenience of their interconnection with the other instrument electronics. The interface circuit is the direct connection between the instrument and the IDPU. It decodes and generates the counter load commands for the preamp boards, and consolidates the serial output telemetry data stream. It also controls the analog multiplexer for the housekeeping data that are sent to the IDPU. A special data product computed by the interface circuit is the electron energy flux. This quantity is derived from the despun field-aligned EESA anode, and is a real-time estimate of the energy flux within a selected energy range that is used by the IDPU trigger algorithms to determine mode changes and burst collection.

The other major function of the interface circuit is to generate a number of analog voltages used to control the high voltage converters and the threshold settings of the MCP pulse amplifiers. All four of the FPGA circuits for the interface and HV control functions have an accurately regulated 5 V supply so they can drive resistor networks for D/A conversion. These circuits generate the six reference voltages that control the high voltage converters and the four threshold control voltages for the MCP pulse amplifiers.

The high voltage sweep generation circuits include waveform synthesis logic and D/A conversion that are contained in two FPGAs, and six independent high voltage drive circuits that use opto-coupled push-pull output stages. The opto-couplers are the AMPTEK HV601. The high output capability of these amplifiers produces a full-scale settling time of about 1 ms for a 4 kV step. The driver circuits for the opto-couplers have a high peak output current capacity that exceeds their continuous dissipation limit. A protection circuit verifies that the raw high voltage supply is greater than a minimum level that will guarantee closed-loop operation of the sweep circuit. IDPU software also checks that the voltage of the fixed HV supply is adequate for the programmed sweep levels.

The high voltage waveform synthesis can generate a total of 174 distinct values. The typical 48 level spectrum is determined by the detector accumulation interval. During each data accumulation the sweep high voltage supply is stepping through 4 micro-step levels. At very low voltages the step resolution only allows one or two micro-steps in each accumulation. The micro-steps are used to approximate a smooth voltage sweep rather than distinct steps. The digital codes for these levels are computed with FPGA hardware logic. Similar synthesis logic is used for both the continuous sweep used by the ESAs and the more complicated sweeps possible with the SESA.

Three of the high voltage outputs serve the SESA sensors, and the other three are used by the I/EESA analyzer and its two deflectors. The deflectors only produce positive deflections to avoid photo-emission, so one of the two deflectors is always at ground potential and the other is sweeping. The deflector voltage is computed to be a fixed fraction of the analyzer voltage step, depending on the desired deflection angle. The angle can be selected from 8 angle levels that are approximately 1.2° per step.

2.4.4. *HV and LV Supplies*

Four power converters supply high voltage to the MCPs, and two additional converters supply the fixed high voltage needed by the sweep generator circuits. All supplies use a sine-wave power oscillator driving a voltage multiplier stack and operate at a frequency of 60 kHz. The nominal maximum voltages are 3 kV for the MCP bias supply voltages and 5 kV for the sweep supply. All converters are voltage programmed and have output current and voltage monitors. Their input supply voltage is 28 V.

The low voltage converters for the stacks are sine-wave converters operating at 60 kHz operate from a regulated 28 V input. There is no output regulation. The primary regulation is supplied by the central power system in the IDPU box.

2.5. TESTING AND CALIBRATION

2.5.1. *Component Testing*

The first level of component testing included qualification of new component types. All electronic components for FAST had to be qualified to a radiation level of 30 kRad or greater. A number of new components that did not have radiation specifications were tested for FAST at the Goddard Space Flight Center (GSFC) test facility. The AMPTEK A121 pulse amplifier the AMPTEK HV601 high voltage opto-coupler were key components for the plasma instrument that did not have radiation data. They were tested and found fully functional after a dose of 100 kRad. The Actel 1020A programmable gate array (FPGA) was used to implement virtually all of the digital logic for the FAST science instruments. This component type had previous test data, and lot testing of the FAST parts found them functional up to 100 kRad.

Functional testing and matching were performed on several key components and subassemblies before installation in circuit boards. All A121 pulse amplifiers were individually tested to determine threshold and gain uniformity, noise susceptibility, and variation of pulse width and dead time with pulse amplitude. The Actel FPGAs were all verified in breadboards prior to assembly.

Microchannel plates underwent an extensive process of preparation and testing. They were inspected, given an ultrasonic alcohol scrub, and baked, prior to vacuum testing. Pairs of plates were initially matched according to manufacturers data on plate resistance and gain, and then were evaluated in vacuum chamber tests. Bell-shaped pulse height distributions with a FWHM of less than 100% were obtained at an electron gain of $\sim 10^6$ for all flight plates. After final selection, MCPs were installed in their flight anode boards, re-tested, and stored in N_2 dry boxes. A total of 20 anode board assemblies (16 flight and 4 flight spares) were assembled and tested.

The analyzer assemblies followed similar test and inspection procedures. Analyzer hemispheres were visually inspected after gold-blacking to check coating uniformity and remove any dust or other contaminants. The analyzer mechanical assemblies underwent preliminary vacuum chamber testing to determine their energy-angle response, azimuthal uniformity, and UV rejection, and to identify any high voltage arcing or other problems. Upon completion of these tests the analyzers were joined with the tested MCP/anode assemblies and the detector electronics boards. The analyzer units were then ready for calibration.

2.5.2. *Calibration*

Calibration and verification of the ESA sensors were performed in three phases. First, a complete calibration and thermal vacuum test was performed at UC Berkeley prior to delivery to GSFC for spacecraft level integration and testing. Second, during the spacecraft thermal vacuum test at GSFC, the analyzer covers were opened and the instruments were run during flight simulations, and stimulated by electron and ion guns and radioactive sources. After completion of the spacecraft integration, the instruments were returned to UCB for inspection and refurbishment. MCPs are sensitive to contamination so there was concern about contamination occurring during the spacecraft tests. The analyzers were opened up for inspection and tested for any change of gain or other symptoms of contamination. No evidence of contamination was found, so the detectors were reassembled for a complete re-calibration and short thermal vacuum test. The entire procedure was completed for all 16 detectors during a two week period. The modular detector design made it possible to disassemble a four analyzer instrument stack down to the individual component level in about four hours. The automated calibration procedure performed a complete angle/energy calibration of an instrument stack in less than one day. Our approach to serviceability and automated testing was fortunate, since launch vehicle delays forced us to repeat this inspection and calibration procedure two more times during the subsequent two years until launch.

The automated ESA calibration used a computer to control a three axis manipulator and high voltage particle accelerators, and also commanded an interactive GSE program on a second computer that controlled the instrument. The electron gun is a photoelectron source that uses a back-illuminated quartz window with a semi-transparent chromium cathode. A grounded planar grid in front of the cathode produces a uniform, parallel electron beam with energies up to 30 keV over a diameter of 10 cm. Helmholtz coils surrounding the chamber were used to null the Earth's field for low energy testing. The ion gun uses an electron impact ionization source. The manipulator has three degrees of freedom: a rotisserie for rotating the sensor around the pitch-angle axis, a yaw axis to change the particle elevation angle, and a linear translation axis that allowed consecutive testing of all four analyzers in the instrument stack. The GSE computer contained an interface card that duplicated the IDPU interface and allowed the computer to configure and read data from the ESA stack. The GSE computer logged all calibration data during the tests.

The ESA calibration procedure included six basic tests:

(1) A complete array of analyzer voltage versus elevation angle data was collected at azimuth angles of $-80°$, $0°$ and $+80°$ for two separate gun energies to determine the analyzer constant (energy/voltage ratio) and verify a uniform energy/angle response.

(2) Hemisphere concentricity was tested by performing an energy sweep at the central elevation angle for each $5°$ azimuth angle step to verify that there was no shift of the detected energy peak with angle.

(3) The optimum MCP bias voltage was found by stepping the programmable threshold of the A121 amplifier to measure the integrated pulse height distribution of the MCP for several values of high voltage bias. This test was made for several values of azimuth angle to map the gain uniformity at different positions on the MCP. This test is also run periodically during flight to monitor MCP gain changes.

(4) Verification of the programmable sweep voltage was made by setting the gun to the nominal ESA energy levels and stepping the hemisphere voltage through equivalent voltage steps.

(5) The relative geometric factor of the anodes within each electron sensor was determined using a Ni^{63} beta source to illuminate the analyzer while the sensor was rotated azimuthally.

(6) The sensor was configured to operate in several flight modes, with high voltage sweeping, while the gun was adjusted to verify that counts were in the proper energy bins.

Individual test runs required between 15 and 60 min to complete. Results from each test were immediately plotted and checked during the subsequent run so any anomalous results could be rechecked.

Absolute geometric factors for the analyzers were estimated from computer simulations, the known grid transparencies, and published values of MCP efficiency. Calibration data from the Ni^{63} beta source test showed $<20\%$ variations in geometric factor around the analyzer azimuth field of view. Sensors were also tested for in-flight uniformity, using relatively isotropic fluxes of plasmasheet particles. Small corrections in sensitivity based on consistent variations in these measurements have been included in data analysis calibrations. The absolute geometric factor for the electron ESAs were tested in-flight by comparing measured magnetic deflection while crossing inverted V arcs with the expected deflection due to current carried by the measured electron flux. These comparisons showed remarkably good agreement (see McFadden *et al.*, 1998b, Figure 1). The electron sensor geometric factor was also checked by comparing the density determined by the ESA data with the absolute density estimated from measured plasma wave cutoffs (Strangeway *et al.*, 1998).

The ion sensor absolute calibration was determined through cross calibration with the electron sensor in the auroral acceleration region. This cross calibration was performed in regions where the energetic (>100 eV) electron and energetic (>30 eV) ion densities dominated the total density. Selecting a region where one could neglect any cold electron component was necessary since photoelectrons produced by the spacecraft and secondary electrons produced within the ESA analyzer can contaminate the low-energy electron measurement. Events were picked where densities could be compared with measured plasma wave cutoffs. The neglect of any low energy ions introduced a small (~20%) correction due to scattered energetic electrons contaminating the low energy ion measurement. This latter assumption was justified by comparison with the low energy ion measurements made by the TEAMS mass spectrometer which was not susceptible to this contamination. In addition, the mass spectrometer data were used to correct for ion composition, although this correction was small because the selected data set had a density dominated by H^+. The ion sensor absolute geometric factor was subsequently reduced by 35% from the estimated pre-flight value in order to obtain agreement between the measured densities. These ion and electron relative geometric factors were subsequently verified on 10 other orbits containing ion beams (McFadden *et al.*, 1999).

The on-orbit performance of the ESA sensors has been exceptional. No radiation damage is evident in the electronics after 3 years of operation. MCP high voltages were increased only during the first year, typically by 100 V, to correct for expected gain decreases during early aging of MCPs. The detectors continue to function normally with low background rates. A discussion of in-flight background rates, including penetrating radiation, solar UV, spacecraft photoelectrons and scattered electrons are discussed by McFadden *et al.* (1999). In summary, the uniformity and stability of the ESA sensors and their ability to accurately obtain absolute measurements of particle fluxes, densities and currents have exceeded expectations.

2.6. IDPU FUNCTIONS AND OPERATING MODES

The IDPU provides command and control of all the instruments on FAST. Since the FAST instruments are shut down each orbit to conserve power, the IDPU must perform a complete initialization of these sensors for each orbit. It was desired that this initialization should be flexible enough to change from orbit to orbit, or even within an orbit. The command load required to accomplish this task was much too large to be uploaded from the ground. Instead, a set of operating modes were developed which would be kept onboard and called out by single commands. This approach reduced the instrument command load to fewer than a dozen commands per orbit, which was compatible with the spacecraft command memory. This small command set includes the particle sensor and field instrument operational mode commands, burst memory trigger commands, fast survey memory trigger commands, and fast survey memory allocation commands. The operational mode command contains a two byte code that selects one of eight stored 'particle mode' commands and one of eight 'field mode' commands. The particle mode configures the ESA and TEAMS instruments and includes command subsets for switching between 'fast survey' and 'slow survey' data collection. Detailed descriptions of instrument commanding are given by Harvey *et al.* (this issue) and McFadden *et al.* (this issue).

The particle mode is used to configure the ESA instrument, including powering low voltage supplies, initializing low voltage electronics, ramping up high voltages, monitoring the currents, setting high voltage sweep waveforms, and initializing the data collection and averaging hardware that generates burst and survey data. One of the eight particle mode commands is reserved for shutting down the ESAs in an orderly sequence at the end of each pass. The particle modes consist of a relatively small command set (<30 commands), but includes some general commands that initiate macro command sets that are common to all the particle modes. Two examples of the macro commands are the 'low voltage on' command which initiates a sequence of ~ 50 commands that initializes the low voltage power supplies and sets instrument control registers, and the 'high voltage on' command that ramps up all the high voltages to their nominal values. Furthermore, these macro commands are controlled by a programmable table which contains default values for all high voltage and low voltage registers. Additional commands in the particle mode are used to control data sampling times and high voltage sweep waveforms as described below. The FAST command strategy provides a relatively flexible but simple method to control the instruments.

The use of various particle modes is primarily to adjust data sampling rates and high voltage sweep waveforms. These functions are controlled through 4 commands: ACCUM, IMODE, EMODE, and SMODE. The ACCUM command controls the data collection time within the ESA, with choices of 1.63, 3.3, or 6.5 ms accumulations. All data collection has been at the highest setting (1.63 ms) since launch and other accumulation times will only be used as a backup if reduced data rates are required.

The IMODE and EMODE commands allow selection of various energy sweeps for the EESA and IESA pairs. Since no instrument degradation has been observed and no science consideration has indicated a need to change this mode, these instruments continue to operate in the normal mode with a 48 step sweep over the full dynamic range each 78 ms. However, the IMODE and EMODE commands allow adjustment of the highest energy sampled or can be set to produce 96 energy step mode with half the time resolution. In case of failure of the EESA, backup commands are available to swap the EESA pair with one of the SESA pairs. The SMODE commands configure the six analyzer pairs that make up the electron spectrograph measurement. The SESAs can be configured into four operating modes, (FIXED, STEPPED, TRACKING, and SWEEPING) discussed below.

The SESA FIXED mode provides the highest time resolution continuous monitor of the pitch angle distribution at 6 energies. This mode is primarily used to monitor electron flux modulations associated with ion cyclotron waves and solitary waves, which are resolved at frequencies up to ~300 Hz (see McFadden *et al.*, 1998b). Factor of 2 separation between energy levels is normally chosen in this mode to provide a wide enough energy range so that samples near the energy flux peak are obtained. For nightside aurora, a 0.5 to 16 keV energy range is normally selected and for dayside aurora the range is reduced to 0.12 to 4 keV.

The SESA STEPPED mode was devised to increase the energy resolution of the measurement at the expense of time resolution. Each SESA pair is stepped through 2, 4 or 8 energy steps, providing a 12, 24 or 48 step energy measurement. The 12 energy sample mode was initially expected to be the best compromise for the SESA since it could provide a wide enough energy range with adequate ($dE/E \sim 1.4$) energy resolution to resolve most features of the distribution. However, it was found that its sampling rate was often inadequate for electron modulations and that its energy resolution was not satisfactory for resolving changes in spectra associated with electrostatic shocks. The 4 and 8 energy step modes have proven to be the most useful for resolving the electron distributions across small scale (~100 ms) electrostatic shocks, providing evidence that these are static structures and not temporal variations in the electric field.

The SESA TRACKING mode was designed to provide a high time and energy resolution measurement of the auroral electron beam. It was known from sounding rocket measurements that higher frequency electron modulations are normally only observed on a steep slope of the distribution above the energy flux peak. Although the FIXED mode with appropriate choice of energy levels normally provides a sample on this steep slope, typically only a single energy level measures statistically significant modulations. The TRACKING mode provides multiple energy samples of electron modulations by selecting 6 closely spaced energy levels that are determined from the electron energy spectra. The IDPU uses measurements from the EESA to determine the energy with the peak count rate. The IDPU then commands the SESAs to a set of energy steps relative to this energy, with levels normally chosen to straddle the energy flux peak. This mode has been useful in measuring small rapid shifts in the spectral peak, which produce out-of-phase changes in count rate above and below the spectral peak, and thus demonstrating that these are modulations of the accelerating electric fields and not modulations of the input electron flux.

The SESA SWEEPING mode was devised to provide a higher time resolution measurement that was identical the EESA measurement. All six SESA pairs are swept over the entire energy range, but their sweeps are phased relative to each other so that the the combined measurement has 6 times the time resolution of the EESA. This mode produces a measurement similar to the STEPPED mode with 48 energies, but reduces problems associated with the high voltage slew rate to a single energy sample. This mode is also useful for cross calibration since all sensors perform the same measurement.

3. Initial Science Results

The high resolution plasma experiment on the FAST satellite has performed as intended and has yielded a wealth of new scientific results. The initial results can be found in the special FAST section of *Geophysical Research Letters* (1998, Vol. 25, number 12). Some highlights of these results include the observation that accelerated ionospheric electrons are a major source of the downward auroral currents, the quantitative demonstration that auroral acceleration is consistent with quasi-static potential acceleration in both the upward and downward current regions, the identification of electromagnetic ion cyclotron (EMIC) wave acceleration as a major modification to the quasi-static potential acceleration, the measurement of electron modulations by electron solitary waves, demonstration that the auroral density cavity is devoid of cold plasma, and conclusive evidence that electron 'horse-shoe' velocity distributions are the energy source for auroral kilometric radiation (AKR). Most of these new observations can be attributed to the design approach outlined in this paper, which combined high energy, angle and time resolution measurements of electron and ion pitch angle distributions with onboard triggers to capture the highest quality observations.

Figure 8 presents data from a nightside auroral crossing that illustrates the range of energy, angle, and time scales covered by the plasma instrument. The spectrograms have not been smoothed so that the measured time resolution can be fully appreciated. The upper two panels show about 5 min of electron data out of a $\sim$35 min data set collected across the northern auroral zone. Observations prior to 10:13 UT were made in a 'fast survey' mode with 312 ms time resolution, whereas the later observations were in 'slow survey' mode with the 2.5 s resolution. The survey mode transition at 10:13 was initiated by the onboard data collection routines to prevent memory overflow.

Several different auroral phenomena can be discerned from the survey data. The top panel shows two large inverted V arcs (10:10:55–10:12:18 and 10:12:30 –10:13:30) that comprise the bulk of the region one upward current system. The dropout in precipitation between them corresponds to a downward current region. The brief appearance of low energy ($\sim$10 eV) upgoing field-aligned electrons, identified by their 180° pitch angle in panel 2, indicates that FAST was crossing near the lower edge of the downward current acceleration region. A much clearer example of accelerated electrons carrying a downward current is seen between 10:10:30 and 10:10:55, where a series of short duration sheets of upgoing elec-

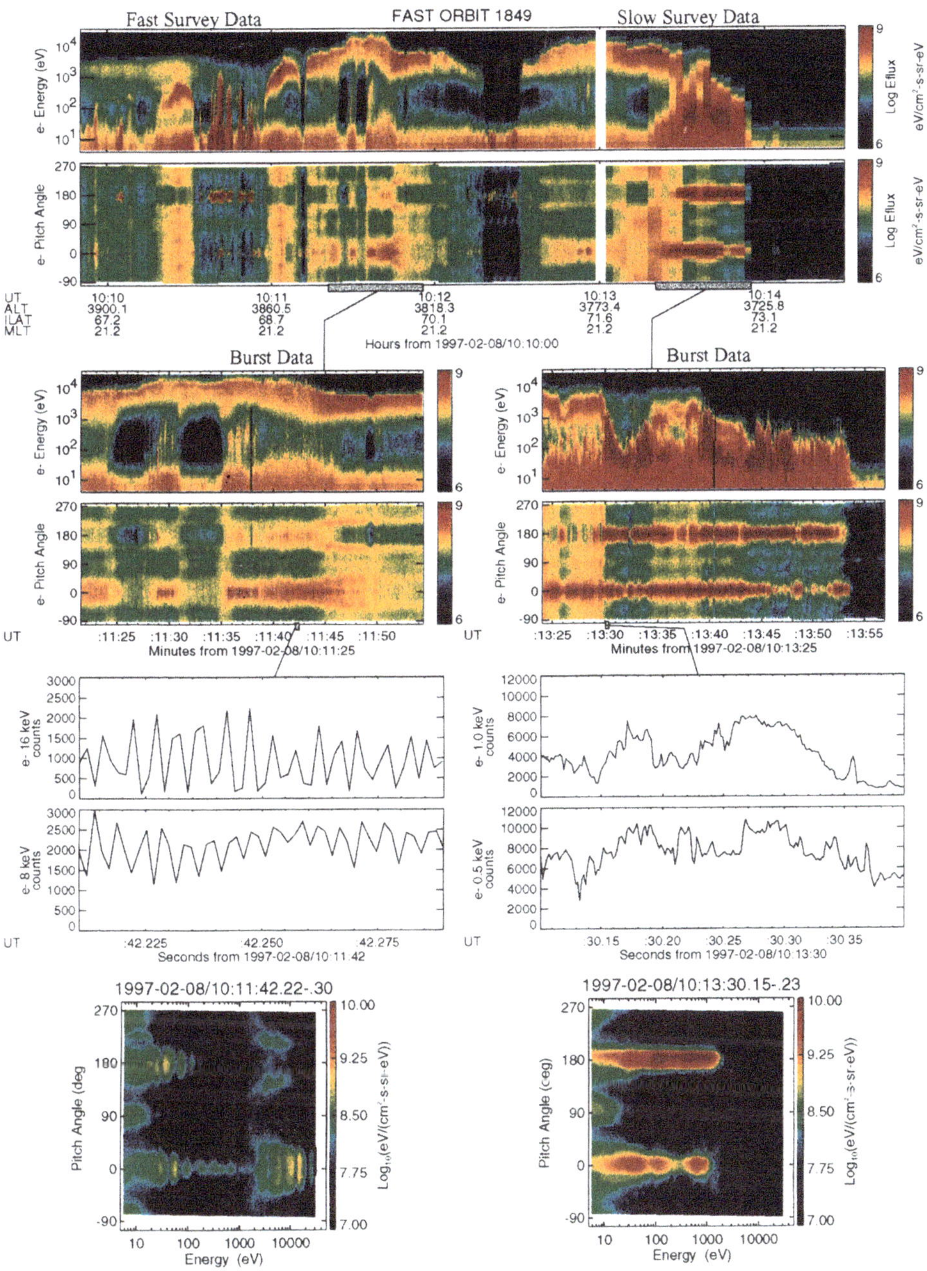

Figure 8. Electron measurements by the FAST satellite during a crossing of the nightside auroral zone. The top two panels show electron energy and pitch angle spectrograms with 'fast survey' and 'slow survey' data appearing before and after 10:13 UT, respectively. Panels 3 and 4 show burst data captured by the onboard triggering system that was keyed to the measured electron energy flux. Panels 5 and 6 show burst data from fixed energy channels of the SESA detectors that show modulations on the 1.6 ms sampling time scale. The bottom two plots are energy-angle distributions of energy flux measured by the EESA illustrating intense field-aligned beams.

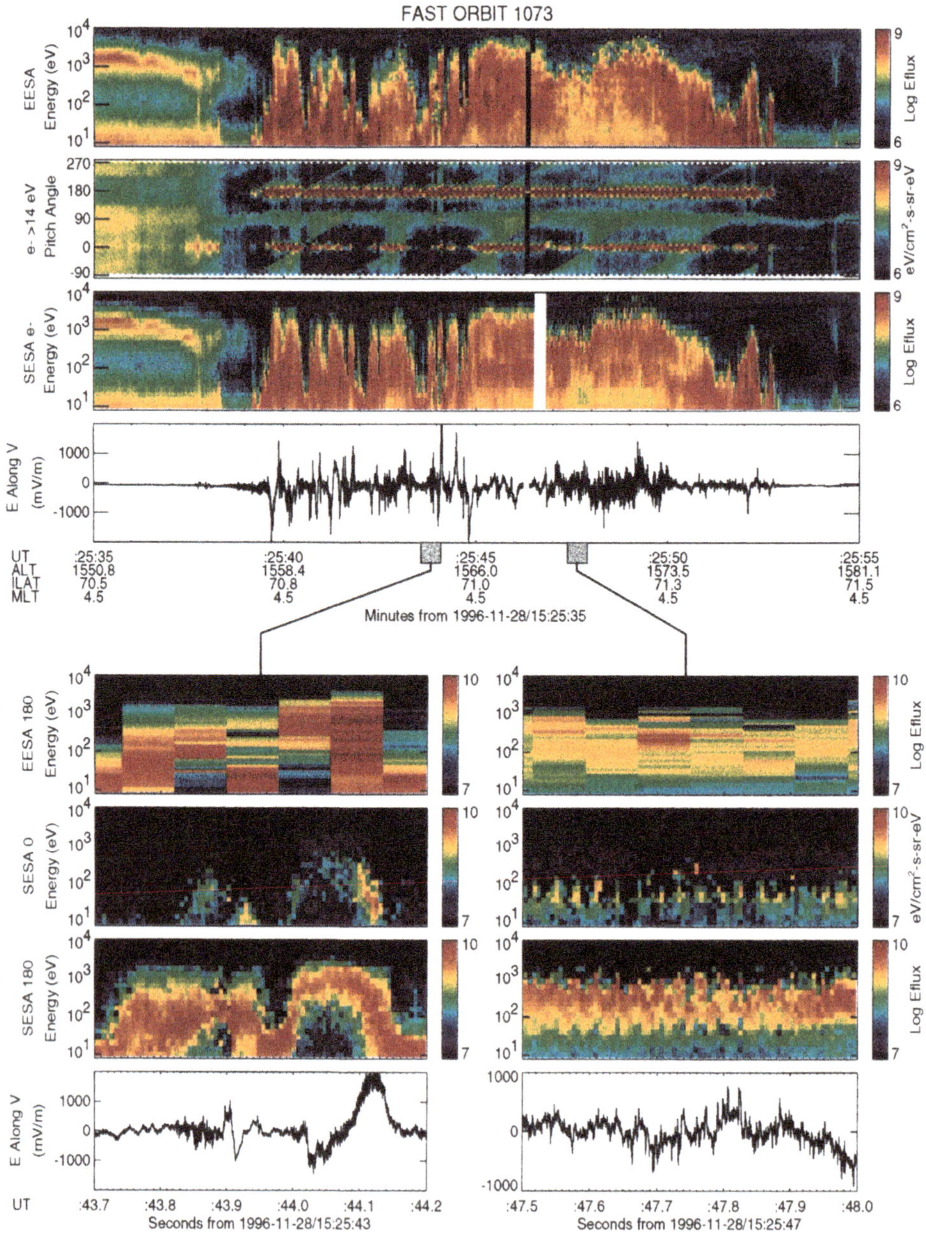

Figure 9. Electron measurements by FAST near the pre-midnight polar cap boundary. These data illustrate some of the physical processes that can be resolved by the high time resolution measurements of the SESA instrument.

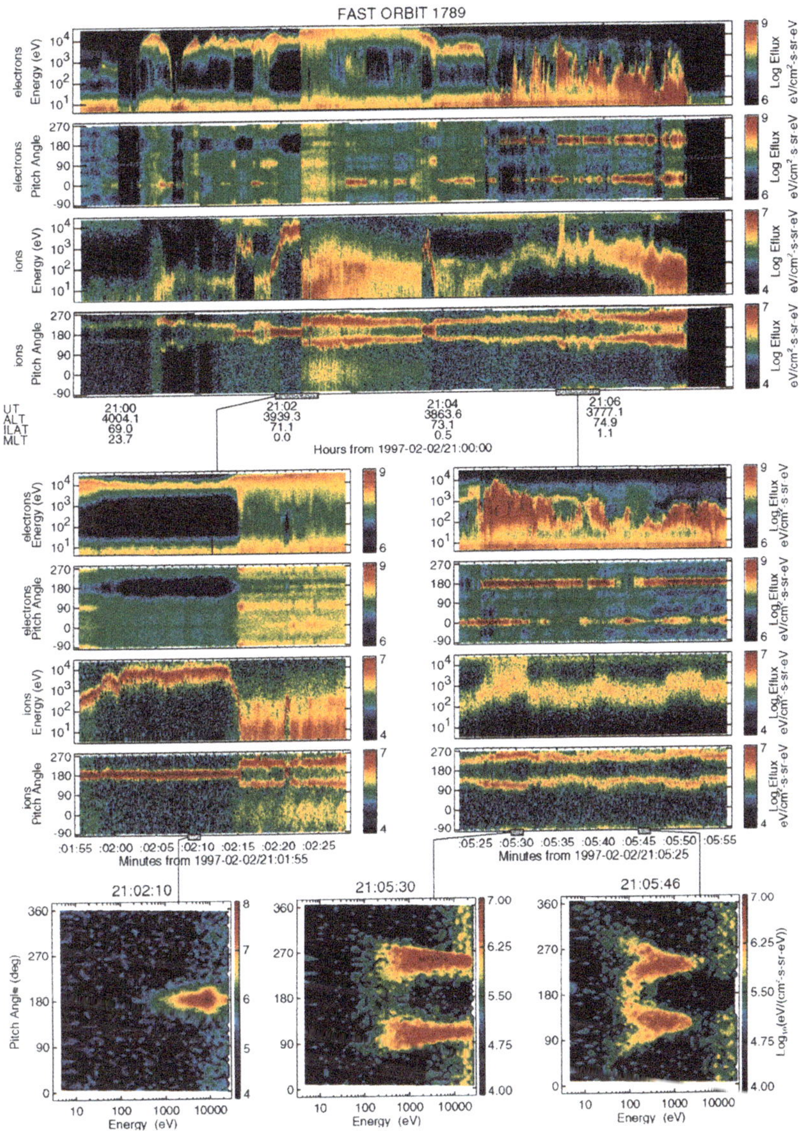

Figure 10. Electron and ion measurements from the FAST satellite during a pre-midnight auroral zone crossing. The configuration of the FAST instruments allow continuous measurements of ion beams and ion conics on time scales much faster than the spacecraft spin period.

trons are observed. The small scale spatial structure of these electrons, coupled with their extremely narrow (~1°) field-aligned pitch angles, made these particles difficult to observe on previous auroral satellites. FAST's continuous monitoring of electron fluxes both upward and downward along the magnetic field, coupled with its high time resolution, has allowed it to observe these electrons on most orbits with apogee in the winter hemisphere indicating they are a common feature that was not obvious prior to FAST.

Two burst captures were triggered during this pass that are displayed with expanded time scales in panels 3 and 4. Their location and duration are identified by the shaded bars at the bottom of panel 2. The burst data reveal structure not resolved in the survey data, such as the short duration bursts of field aligned electrons at energies visible below the inverted V spectral peak in the first burst. These field aligned electrons are cold electrons that have been accelerated by EMIC waves. Rapid modulations of the electron distribution caused by these waves were captured by the 8 and 16 keV fixed energy channels of the SESA as shown in the expanded time scale panels 5 and 6 (left side). These 1.6 ms measurement samples are just able to resolve the $\sim$220 Hz modulations that occur just below the H^+ cyclotron frequency. These modulations also appear in the energy-angle distributions where they cause intensity modulations at different energies during the energy sweep (left side, bottom panel). The most intense modulations by EMIC waves are observed within inverted V arcs that include a lower temperature, magnetic field-aligned electron population (left side, panels 4 and 7).

Another feature often seen in the FAST data is counterstreaming electrons at the polar cap boundary near midnight (10:13:25–10:13:55 UT). The large electron flux in this region triggered burst data capture of the event as seen in panels 3 and 4 (right side). These EESA burst data, with 78 ms time resolution, capture a wealth of structure in the electrons completely missing from the survey data. However, even these high resolution data fail to capture all the structure. Panels 5 and 6 (right side) show the 1.6 ms time resolution data from two fixed energy channels of the SESA. Structure in the precipitation is seen down to the smallest time scales. This structure will often reveal itself in the distribution function measurements (right side, bottom panel) as time aliasing in the detector energy sweep. The bottom panel displays a counterstreaming electron distribution illustrating the broad energy range over which this precipitation is observed. The pitch angle width of the beam in the plot is limited by the detector resolution (11.2°). However, the use of fixed energy SESAs has allowed an estimate of the beam width by examining the rise in count rate as fixed energy detectors rotate through these beams. These measurements establish that the beams are extremely narrow ($\sim$1°) with perpendicular temperatures comparable to the ionospheric thermal population.

The counterstreaming and upgoing electron beams often contain rapid time variations in their energy structure that are best resolved by the SESA cycle mode. Figure 9 shows another example of the polar cap boundary region. The top two panels present a 20 second burst capture of electron energy flux and pitch-angle spectra collected by the EESA, with 78 ms time resolution. These data have resolution comparable to the burst example in panels 3 and 4 of Figure 8. The third panel displays SESA burst data with 24 energy levels and 6.5 ms time samples that resolve time structure missing in the EESA data. The character of the electron fine structure changed about half-way through this event, which is also evident in the electric field data plotted in panel 4. The first half of the event contains structured upgoing electron beams that are correlated with large bipolar electric field structures. These field signatures are consistent with quasi-static U-shaped potential structures that accelerate the electron beams. The second half of the event contains higher frequency electric field variations and fine-scale modulation of the electrons that result from wave-particle interactions in the upgoing electron beam.

The two sets of panels in the lower half of Figure 9 are expanded plots of 0.5 second intervals from these two regions. Panel 5 displays the upgoing (180°) magnetic-field-aligned electron flux measured by the EESA, which can be compared with the upgoing (180°) and downgoing (0°) fluxes measured by the SESA shown in panels 6 and 7. The panels on the left show a cut across two distinct accelerated beams. The second beam, which is crossed in about 0.1 s starting at 15:25:44.0, is the more energetic and distinct event. The bipolar electric field signature in the bottom panel has an amplitude greater than 1 volt m^{-1}. The EESA measurement barely resolves this structure, whereas it is resolved in detail by the SESA. Assuming that this beam is purely a spatial feature, it maps to a 0.6 km width at ionospheric heights, and the individual data samples represent a 40 m scale size. The right-hand set of panels 5–8 show a different signature of electron energy variation, which is the rapid energy modulation of the upgoing electron beam. In this case, the only evidence recorded by the EESA is the aliased modulation during the energy sweep. The SESA data show a quasi-periodic modulation of the electron energy by several hundred volts at approximately 30 Hz. The electric field contains intense broad-band low frequency waves with an amplitude of several hundred mV/m. The electron modulation is apparently caused by interaction with these intense ion cyclotron and fast solitary waves.

Figure 10 highlights some of the ion measurements made on a typical midnight sector auroral pass. The top four panels shown 8 minutes of electron and ion fast survey data near the high latitude boundary of the auroral zone during a moderately active period (AE $\sim$ 500). The electron measurements in the first two panels show a signature similar to the event in Figure 8. This includes a wide region of inverted V electron acceleration between 21:00:00 and 21:04:30 UT, followed by an interval of upgoing and counterstreaming electron beams near the polar cap boundary. Panels 3 and 4 show typical ion properties of this region. Three ion beams are easily identified in panel 3 by their intense energy flux and narrow energy spread. The pitch angle distributions in panel 4 show that the beams are closely aligned near 180°.

Conical ion distributions are evident by their pitch-angle signature, which lies near the loss cone angle on both sides of 180°. Conics within the inverted V have broad energy spectra typically ranging from a few eV to a few hundred eV. The ion conic signature is distinctly different in the region of upgoing electron beams near the polar cap boundary. Here the ion conics are generally more energetic, with spikes of high energy coinciding with the most energetic electron beams. The most intense, although less energetic conic fluxes are found near the polar cap boundary, coincident with the intense counterstreaming electron fluxes.

The left set of burst panels (5–8) display a high resolution view of the ion beams and conics found in the inverted V regions. Panels 7 and 8 demonstrate the ion beam energy variations and the sharp boundaries typical of ion beams. A low energy ion beam lasting only a fraction of a second is fully resolved in the ion energy flux and pitch angle plots at 21:02:21 UT. The ion beams are created when a portion of the inverted V electrical potential extends below the altitude of the spacecraft and accelerates ionospheric ions upward. The potential above the spacecraft decreases by the same amount, which is apparent from the slight decrease of the electron beam energy coincident with the ion beams. The potential below the spacecraft also suppresses ionospheric thermal and secondary electrons and produces a density cavity within the ion beam, as is evident in panel 5. The complete depletion of the electron loss cone is also clear from panel 6. A typical pitch angle distribution for the beam ions is found in the bottom left panel. The fine-scale structure within the beam results from slight differences between energy distributions for each of the ion species that arises from streaming interactions between them.

Ion and electron distributions change dramatically as we move toward the polar cap boundary. The ion conic heating is much more intense in regions of upgoing electron beams. The burst panels on the right show data from a region that includes electron solitary waves as well as energetic electron beams and energetic conics. The pitch angle plot at the bottom center is a 1.5 s average of ion data from the center of the most intense heating event at 21:05:30 UT. The peak energy of the ions extends to 20 keV, and the pitch angles lie generally between the ionospheric loss cone and 90°, with the peak energy at intermediate angles. Another conic pitch angle plot in the lower right box is a period of somewhet less intense heating. Both of these conic events display a rather high minimum energy which suggests that they are trapped by the potential that is accelerating the electron beams. The ions can only escape when they have gained sufficient ‘perpendicular momentum’ for their magnetic mirror force to overcome the parallel electric field.

4. Summary

The FAST electron and ion plasma analyzers were designed to measure auroral particle velocity distributions with an unprecendented combination of time, energy, and pitch angle resolution. Data are selected by on-board evaluation of the measurements to make the best use of the available telemetry rate. Measurements during these burst events can resolve ion and electron variations on ion cyclotron time scales and spatial scales of the electron inertial length. The on-orbit performance of the instrument has met all expectations.

Acknowledgements

The authors are grateful to the many people who contributed to the success of this project. We give special recognition to Roland Scholz, who played a major role in assembling the FAST hardware, and was also instrumental in building many of the rocket and satellite experiments that led up to this mission.

References

Boehm, M., Paschmann, G., Clemmons, J., Hoefner, H., Frenzen, R., Ertl, M., Haerendel, G., Hill, P., Lauche, H., Eliasson, L. and Lundin, R.: 1994, 'The TESP Electron Spectrometer and Correlator (F7) on Freja', *Space Sci. Rev.* **70**, 509.

Carlson, C. W., Curtis, D. W., Paschmann, G., and Michael, W.: 1983, in D. McDiarmid (ed.), 'An Instrument for Rapidly Measuring Plasma Distribution Functions with High Resolution', *Adv. Space Res.* **2**, 67.

Carlson, C. W., Pfaff, R. W., and Watzin, J. G.: 1998a, 'The Fast Auroral Snapshot (FAST) mission', *Geophys. Res. Lett.*, **25**, 2013.

Carlson, C. W. and McFadden, J. P.: 1998, in R. F. Pfaff, J. E. Borovsky, and D. T. Young (eds.), 'Design and Application of Imaging Plasma Instruments', *Measurement Techniques in Space Plasmas: Particles*, Geophys. Monogr. **102**, AGU, Washington, p. 125.

Ergun, R. E., Carlson, C. W., Mozer, F. S., Delory, G. T., Temerin, M., McFadden, J. M., Pankow, D., Abiad, R., Harvey, P., Wilkes, R., Primbsch, H., Elphic, R., Strangeway, R., Pfaff, R., and Cattell, C. A.: 2000, 'The FAST Satellite Electric Field and Magnetic Field Instruments', *Space Sci. Rev.*, this issue.

Gurnett, D. A.: 1974, 'The Earth as a Radio Source: Terrestrial Kilometric Radiation', *J. Geophys. Res.* **79**, 4227.

Harvey, P. R., Curtis, D. W., Heetderks, H. D., Pankow, D., Rauch-Leiba, J. M., Wittenbrock, S. K., and McFadden, J. P.: 2001, 'The FAST Spacecraft Instrument Data Processing Unit', *Space Sci. Rev.*, this issue.

Klumpar, D. M.: 1986, in T. Chang (ed.), 'A Digest and Comprehensive Bibliography on Transverse Auroral Ion Acceleration', *Ion Acceleration in the Magnetosphere and Ionosphere*, Geophys. Monogr. **38**, AGU, Washington, p. 389.

McFadden, J. P., Carlson, C. W., and Boehm, M. H.: 1986, 'Field-aligned Electron Precipitation at the Edge of an Arc', *J. Geophys. Res.* **91**, 1723.

McFadden, J. P., Carlson, C. W., Boehm, J. H., and Hallinan, T. J.: 1987, 'Field-aligned Electron Flux Oscillations that Produce Flickering Aurora', *J. Geophys. Res.* **92**, 11133.

McFadden, J. P. and Carlson, C. W.: 1998, in R. F. Pfaff, J. E. Borovsky, and D. T. Young (eds.), 'Computer Simulation in Designing Electrostatic Optics for Space Plasma Experiments', *Measurement Techniques in Space Plasmas: Particles*, Geophys. Monogr. **102**, AGU, Washington, p. 249.

McFadden, J. P., Carlson, C. W., Ergun, R. E., Chaston, C. C., Mozer, F. S., Temerin, M., Klumpar, D. M., Shelley, E. G., Peterson, W. K., Moebius, E., Kistler, L., Elphic, R., Strangeway, R., Cattell, C., and Pfaff, R.: 1998b, 'Electron modulation and ion cyclotron waves observed by FAST', *Geophys. Res. Lett.* **25**, 2045.

McFadden, J. P., Carlson, C. W., and Ergun, R. E.: 1999, 'Microstructure of the auroral acceleration region as observed by FAST', *J. Geophys. Res.* **104**, 14453.

McFadden, J. P., Ergun, R. E., Carlson, C. W., Herrick, W., Loran, J., Vernetti, J., Teitler, W., Bromund, K., and Quinn, T.: 2001, 'Science Operations and Data Handling for the FAST Satellite', *Space Sci. Rev.*, this issue.

Mozer F. S., Carlson, C. W., Hudson, M. K., Torbert, R. B., Parady, B. Yatteau, J., and Kelley, M. C.: 1977, 'Observation s of paired Electrostatic Shocks in the Polar Magnetosphere', *Phys. Rev. Lett.* **38**, 292.

Pankow, D., Besuner, R., Ullrich, R., and Wilkes, R.: 2001, 'Deployment Mechanisms on the FAST Satellite: Radial Wires, Stiff Axials, and Magnetometer Booms', *Space Sci. Rev.*, this issue.

Pfaff , R. , Carlson, C. W., Watzin, J., Everett, D., and Gruner, T.: 2001, 'An overview of the FAST Auroral Snapshot (FAST) Mission', *Space Sci. Rev.*, this issue.

Sharp, R. D., Shelley, E. G., Johnson, R. G., and Ghielmetti, A. G.: 1980, 'Counterstreaming Electron Beams at Altitudes of ~1 RE over the Auroral Zone', *J. Geophys. Res.* **85**, 92.

Strangeway, R. J., Kepko, L., Elphic, R. C., Carlson, C. W., Ergun, R. E., McFadden, J. P., Peria, W. J., Delory, G. T., Chaston, C. C., Temerin, M., Cattell, C. A., Moebius, E., Kistler, L. M., Klumpar, D. M., Peterson, W. K., Shelley, E. G., and Pfaff, R. F.: 1998, 'FAST observations of VLF waves in the auroral zone: Evidence of very low plasma densities', *Geophys. Res. Lett.* **25**, 2065.

Temerin, M. A., Carlson, C. W., Cattell, C. A., Ergun, R. E., McFadden, J. P., Mozer, F. S., Klumpar, D. M., Peterson, W. K., Shelley, E. G., and Elphic, R. C.: 1990, in T. Chang (ed.), 'Wave-particle Interactions on the FAST Satellite', *Physics of Space Plasmas*, Scientific Publishers, Inc., Cambridge, MA., p. 343.

THE FAST SATELLITE FIELDS INSTRUMENT

R. E. ERGUN*, C. W. CARLSON, F. S. MOZER, G. T. DELORY, M. TEMERIN, J. P. McFADDEN, D. PANKOW, R. ABIAD, P. HARVEY, R. WILKES and H. PRIMBSCH

Space Sciences Laboratory, University of California, Berkeley, CA, U.S.A.

R. ELPHIC

Los Alamos National Laboratory, Los Alamos, NM, U.S.A.

R. STRANGEWAY

University of California, Los Angeles, CA, U.S.A.

R. PFAFF

Goddard Space Flight Center, Greenbelt, MD, U.S.A.

C. A. CATTELL

University of Minnesota, Minneapolis, MN, U.S.A.

Abstract. We describe the electric field sensors and electric and magnetic field signal processing on the FAST (Fast Auroral SnapshoT) satellite. The FAST satellite was designed to make high time resolution observations of particles and electromagnetic fields in the auroral zone to study small-scale plasma interactions in the auroral acceleration region. The DC and AC electric fields are measured with three-axis dipole antennas with 56 m, 8 m, and 5 m baselines. A three-axis flux-gate magnetometer measures the DC magnetic field and a three-axis search coil measures the AC magnetic field. A central signal processing system receives all signals from the electric and magnetic field sensors. Spectral coverage is from DC to ~4 MHz. There are several types of processed data. *Survey data* are continuous over the auroral zone and have full-orbit coverage for fluxgate magnetometer data. *Burst data* include a few minutes of a selected region of the auroral zone at the highest time resolution. A subset of the burst data, *high speed burst memory data*, are waveform data at 2×10^6 sample s^{-1}. Electric field and magnetic field data are primarily waveforms and power spectral density as a function of frequency and time. There are also various types of focused data processing, including cross-spectral analysis, fine-frequency plasma wave tracking, high-frequency polarity measurement, and wave-particle correlations.

1. Introduction

The primary scientific goal of the FAST mission is to investigate auroral electron acceleration, ion heating, and wave-particle interactions (Temerin *et al.*, 1990). These phenomena involve a rich variety of plasma processes which include electrostatic shocks (Mozer *et al.*, 1977), double layers (Temerin *et al.*, 1982), field-

*Currently at the Department of Astrophysical and Planetary Sciences and the Laboratory for Atmospheric and Space Physics, University of Colorado, Boulder, CO, U.S.A.

Space Science Reviews **98:** 67–91, 2001.

aligned electrons (McFadden *et al.*, 1986), Langmuir and whistler wave emissions (Gurnett *et al.*, 1969), auroral kilometric radiation (AKR) (Gurnett, 1974), ion conics (Klumpar, 1986), ion beams, and the formation of the auroral density cavity. The FAST mission was designed to have one to three orders of magnitude higher time resolution than previous auroral satellite missions which have identified many of the auroral processes but were unable to resolve them fully.

The FAST satellite was launched into an ~83° inclination orbit with a 350 km perigee and 4175 km apogee in August, 1996. The satellite is oriented in a 'cart-wheel' attitude which has the spin axis nearly (negative) normal to the orbital plane. It is spin stabilized with a spin period of 5 s. The satellite crosses the auroral zones (which form ovals at ~65°–70° magnetic latitude North and South) four times an orbit. The orbit was designed to have a Northern apogee during January and February of 1997 for coordinated ground-based and optical observations.

In situ observations of auroral processes at high resolution require high data rates from all instruments which would exceed telemetry capability. Because the auroral zones are only a fraction of the FAST orbit, it is possible to acquire data at many times the maximum telemetry rate if the data are stored on-board. Central to the FAST satellite is a large (~128 MByte) solid state memory common to all instruments.

The FAST data system is described elsewhere (Harvey *et al.*, 2001) so we provide only a brief overview. The on-board memory accepts two primary types of science data from all instruments: *survey data* which have continuous coverage over the entire auroral zone and *burst data* which have only a few minutes of coverage (corresponding to ~500 km of ground track) at the highest resolution. The position of the auroral zone, ~65°–70° magnetic latitude, varies enough that on-board triggers are required for the burst data. Survey data has three sub-types: 'full-orbit' data which is typically for the fluxgate (DC) magnetometer only, 'slow survey' which has ~0.06 s resolution waveforms in the sub-auroral and polar cap regions, and 'fast survey' which typically has ~0.5 ms resolution.

The basic plan of operation is as follows. The FAST satellite always acquires magnetic field data. As it approaches auroral latitudes, the science instruments are put into 'slow survey' by a time-tagged command. Once the auroral electron precipitation is detected, the science instruments are configured into 'fast survey'. If signals of scientific interest are seen by the burst triggers, burst data is collected for between ~5 s to several minutes. Collection typically includes 25% of data prior to the trigger. There may be several burst collections depending on the available memory. The survey and burst data are stored on board until telemetered to ground.

The Fields Instrument includes the electric field and magnetic field sensors, the electric and magnetic field signal processing, and a wave-particle correlator. This article will detail the electric field sensors and the electric and magnetic field signal processing system. The magnetic field sensors (Elphic *et al.*, 2001), and the wave-particle correlator (Ergun *et al.*, 1998), and the booms (Pankow *et al.*, 2001) are described in separate articles.

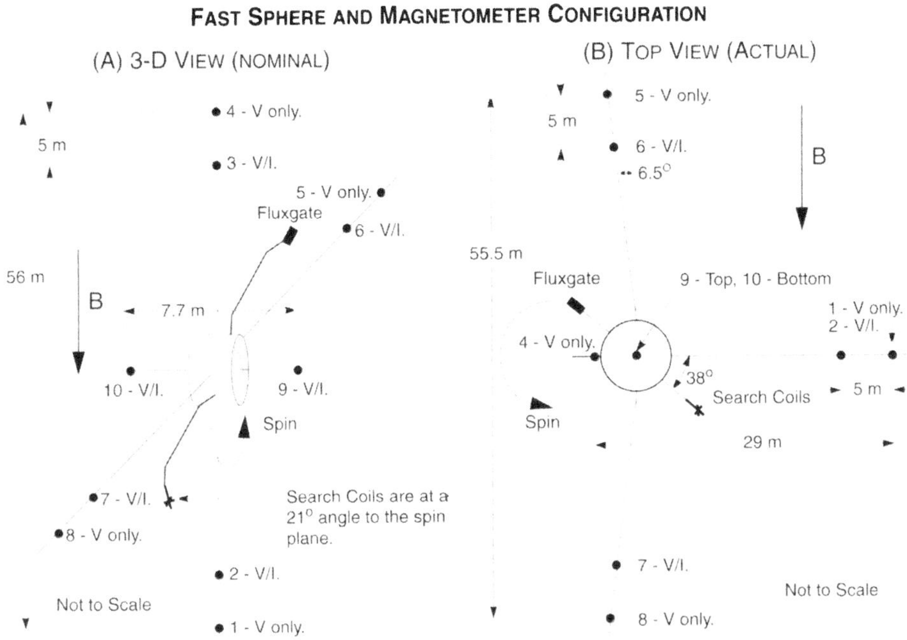

Figure 1. (a) A three-dimensional view of the electric and magnetic field sensors on the FAST satellite as designed. The electric field instrument has eight spherical sensors that are on four wire booms (two each) in the spin plane and two that are on rigid axial booms. All of the spherical sensors can operate in 'voltage mode' (marked with 'V') in which they measure the local plasma potential with respect to the payload. The electric field signals are measured by pairs of sensors which form 56 m, 7.7 m, or 5-m dipoles. Six of the ten sensors, marked with 'I', can operated in current mode where the electron current is measured for deriving plasma density. The fluxgate and search coil magnetometers are on ~2-m booms. The search coil assembly is rotated 21° out of the spin plane. (b) The deployed state. The wire boom carrying sensors 3 and 4 did not fully deploy.

2. Fields Instrument Overview

The Fields Instrument has a number of sensors and signal processing subcomponents that are designed to produce the survey and burst data in accordance with the FAST data system. In this section, we briefly describe subcomponents of the Fields Instrument and show how they are linked together to provide full spectral coverage of electric and magnetic fields. The subcomponents are described in detail in Sections 3–5.

2.1. Sensors

Figure 1 diagrams the electric and magnetic field sensors in their nominal (Figure 1(a)) and actual (Figure 1(b)) positions. There are ten spherical sensors for measuring the electric field. Eight sensors are on four wire booms (one of which did not fully deploy) in the spin plane and two are on rigid booms that are deployed

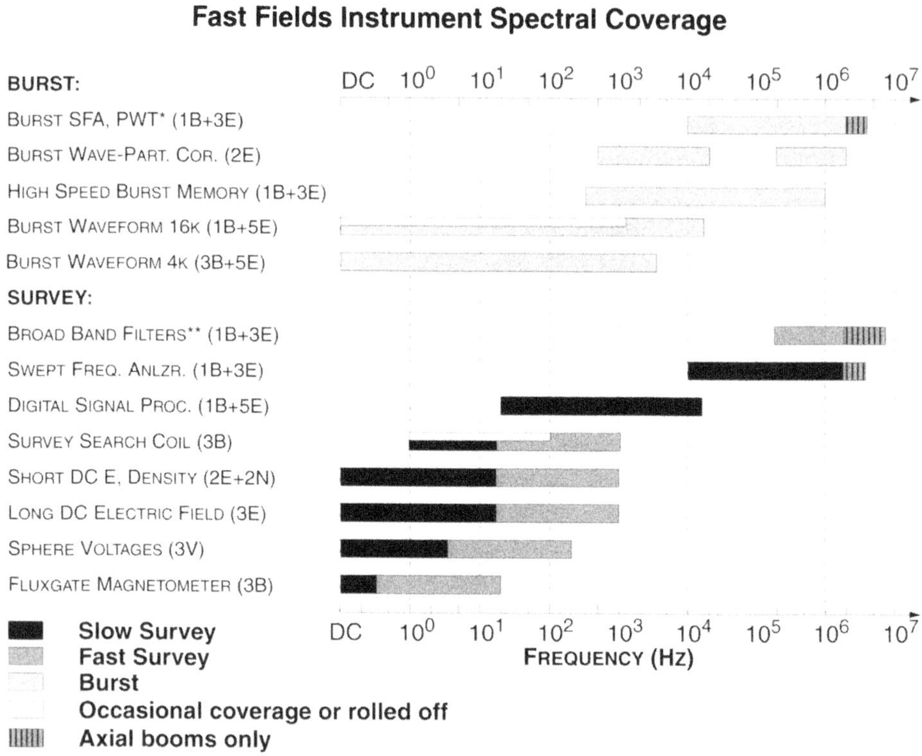

* PWT - Plasma Wave Tracker - Data in Burst Waveform 16k and DSP.
** Broad Band Filters - Includes High Frequency Phase Packet (HFQ).

Figure 2. The spectral coverage of fields instrument.

along the spin axis. The vector DC and AC electric field signals are derived from the voltage difference between pairs of spherical probes which form dipoles of 5 m or 56 m in the spin plane (Mozer, 1973). The spin-plane wire booms have two sensors each to make multi-point measurements which can be used to determine the wave vector of coherent emissions with cross-spectral analysis. The tip-to-tip length of the spin axis dipole is 7.7 m. Six of the ten sensors can operate as Langmuir probes, measuring the electron current to a probe at a fixed potential to determine the electron density.

The DC magnetic field (to ~100 Hz) is measured by a three-axis fluxgate magnetometer and the AC magnetic fields are measured by three search coils (~10 Hz to 4 kHz on two axes, ~10 Hz to 500 kHz on one axis). The fluxgate magnetometer is on a 2-m boom opposite an identical boom which carries the search coils. The satellite is oriented so that the magnetic field lies within 6° of the spin plane in the auroral regions.

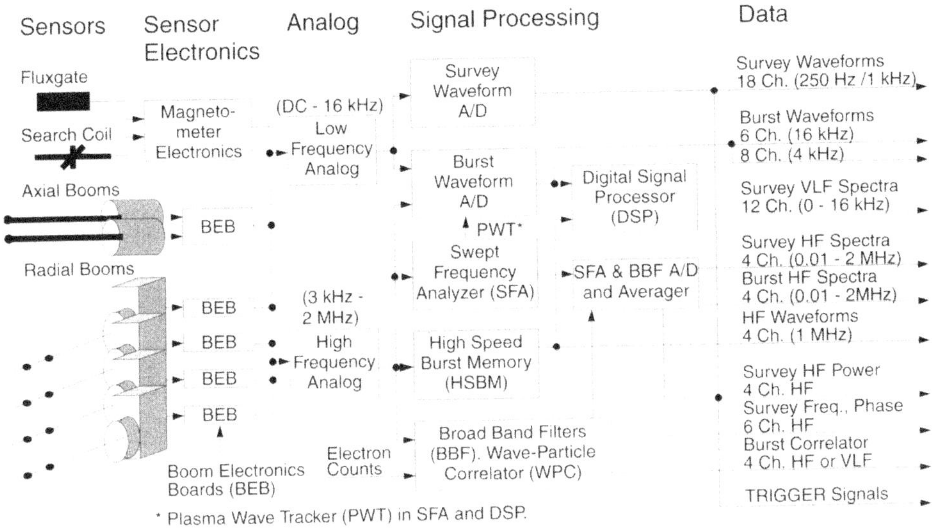

Figure 3. A block diagram of the fields instrument.

2.2. Spectral Coverage

The Fields Instrument was designed to provide full spectral coverage (DC to 2 MHz) of electric and magnetic fields continuously as survey data (Figure 2). The lowest frequency data (up to 1 kHz) are vector waveforms of electric and magnetic fields, and waveforms of plasma density. The DSP (Digital Signal Processor) produces wave spectra from 16 Hz to 16 kHz and the SFA (Swept Frequency Analyzer) produces wave spectra from ~10 kHz to 2 MHz. DSP and SFA spectra have ~4 s resolution in slow survey and typically ~0.25 s resolution in fast survey. The BBF (Broad Band Filters) provide high time resolution amplitude, frequency, and polarization of >200 kHz coherent wave signals in fast survey only.

Burst data (Figure 2) supplement the survey data with 16 kHz Nyquist-frequency waveforms, high time resolution SFA spectra (62.5 ms), PWT (Plasma Wave Tracker), and WPC (Wave-Particle Correlator) data. In addition, the HSBM (High-Speed Burst Memory) records four channels of waveforms, typically three E and one B, with a frequency range from ~1 kHz to 1 MHz.

2.3. Signal Processing

Boom Electronics Board (BEB)

Figure 3 shows a functional overview of the Fields Instrument. The signals from the spherical sensors are passed through a BEB located at the boom deployment unit. The BEB supplies bias currents to the electric field sensors, controls the boom motors during deployment, and provides housekeeping signals to the instrument processor.

Low Frequency Analog and High Frequency Analog

The 16 electric and magnetic field sensors share a common signal processing system that is located in the central instrument electronics box. The signals are fed to two analog conditioning circuits. The low frequency analog conditioning covers the frequency band from DC to ~16 kHz and the high frequency analog conditioning typically covers the frequency range from ~3 kHz to 2 MHz (4 MHz maximum). The analog conditioning circuits have differential amplifiers to form the electric field signals from pairs of sensors and filters to isolate configured frequency bands. Analog switches and analog multiplexors allow the instrument to operate in a variety of modes.

Survey Waveforms

Survey waveforms (Figure 3) include 18 electric and magnetic field signals. The survey waveforms have maximum sample rates of 2048 samples s^{-1} (10 such channels at ~1 kHz Nyquist) and 512 samples s^{-1} (8 such channels ~250 Hz Nyquist).

Burst Waveforms

The burst waveforms unit has 8 A/D converters at 32 768 samples s^{-1} (~16 kHz Nyquist). It operate in one of two modes. One mode has 8 signals at 32 768 samples s^{-1}. The other has 6 signals at 32 768 samples s^{-1} and 8 signals at 8196 samples s^{-1} (~4 kHz Nyquist). Survey and burst waveforms are labeled by the band width: 16 kHz, 4 kHz, 1 kHz, or 250 Hz.

Digital Signal Processor (DSP)

Continuous coverage of the power spectral density of the 16 kHz signals is computed onboard by the DSP (Figure 3) which averages several Fast Fourier Transforms (FFT). The 1024 point FFT covers the frequency range from DC to ~16 kHz with 32 Hz bandwidth and ~100 dB dynamic range. The DSP can also perform a cross-spectral analysis to determine the phase difference of pairs of signals. The spectra produced by the DSP are treated as survey data, giving continuous coverage in the auroral zone.

Swept Frequency Analyzer (SFA)

High-frequency signals (>16 kHz) are processed three ways (Figure 3). The SFA produces power-frequency-time spectra typically from ~10 kHz to 2 MHz (the sweep range is adjustable) with ~80 dB dynamic range and 15 kHz bandwidth. The time resolution is typically 62.5 ms per spectra (31.25 ms is the fastest resolution) which are transmitted as burst data. Averaged spectra (125 ms to 4 s resolution) form survey data. The SFA unit also contains a Plasma Wave Tracker (PWT) which gives fine frequency (~50 Hz) resolution over a narrow frequency range (16 kHz) that lies between 0 and 2 MHz.

High-Speed Burst Memory (HSBM)

The HSBM digitizes four high-frequency signals (~3 kHz–1 MHz) at 2×10^6 samples s^{-1} with 10-bit resolution. HSBM data are stored in 2.5 Mbyte buffers that cover ~0.25 s periods. Data intervals are selected by dedicated triggers which monitor wave power in both the high-frequency (200 kHz–2 MHz) and low-frequency (~1 kHz–16 kHz) bands. The HSBM has very limited time coverage (<1% duty cycle) due to the high data rates.

Broad-Band Filters (BBF)

The BBF rectify four electric or magnetic field signals to determine the amplitude envelope of 200 kHz–2 MHz wave emissions versus time (~1 ms resolution). They have a dynamic range of ~60 dB. For the same four signals, the number of zero crossings are counted, which represents the wave frequency of narrow-band emissions. The phase shift between each pair of the four selected high-frequency signals (six phase difference signals) are also measured on board to determine the high-frequency wave polarization. The high-frequency wave amplitudes, zero crossing rates, and polarizations are telemetered as survey data.

Wave-Particle Correlator

The wave-particle correlator uses two electric or magnetic field signals and twelve of the stepped electron electrostatic analyzer anode signals (Carlson *et al.*, 1998a) to measure oscillations in electron fluxes in one of two selected frequency ranges. The low frequency range is from ~500 Hz to ~16 kHz and the high frequency range is from ~200 kHz to ~2 MHz. The correlation function is computed onboard by digital circuitry.

3. Electric Field Sensors, Booms, Preamplifiers, and Boom Electronics Board

3.1. Antennas and Sensors

The sensors for measuring the electric field are shown in Figure 1. The 8-cm diameter spherical probes contain electronics that operate in one of two selectable modes. All of the sensors can operate in 'voltage mode' in which they measure the potential of the nearby plasma with respect to the spacecraft. In voltage mode the probes are biased with a fixed current. Six of the sensors (marked with 'I' Figure 1) can operate in 'current mode' as Langmuir probes which measure plasma current. In current mode, the probe is biased at a fixed potential.

The radial booms are 2.5 mm diameter wires which support the sensors and carry power and signals between the sensors and the spacecraft. The wires have a kevlar braid that surrounds two coaxial cables and eight insulated wires. The kevlar braid is covered with aluminized kapton and a silver coated copper wire

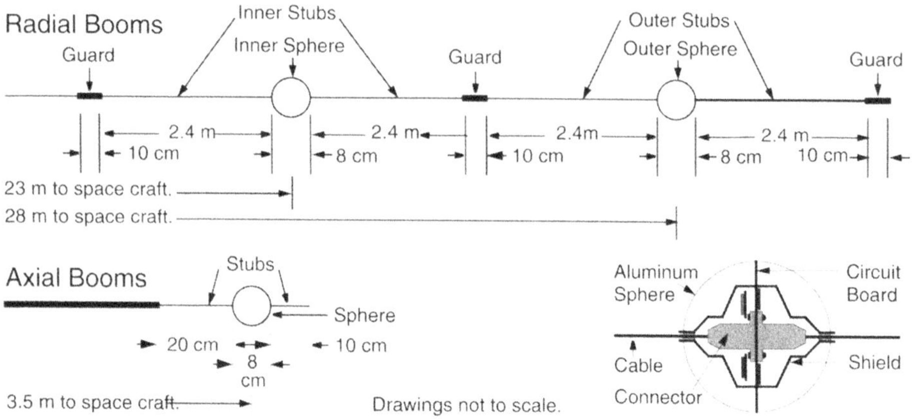

Figure 4. Top: the physical layout of the radial boom spheres, stubs, and guards. There are two spheres on each wire separated by 5 m. The 8-cm diameter spheres house a preamplifier. The outer sphere operates in voltage mode only. The inner sphere (nearer the spacecraft) operates in voltage or current mode. Adjacent to each sphere are 2.4-m stub sections which are biased at a selectable voltage with respect to the sphere potential. Three 10-cm long guard sections are also shown. Bottom left: the axial boom physical layout. There is one sphere on each ~4-m rigid boom. Each sphere has a 20-cm stub on the inside and a 10-cm stub on the outside. The axial spheres can operate in voltage or current mode. Bottom right: the mechanical sphere assembly. Inside of each sphere are one or two circuit boards for voltage mode and/or current mode operation. Surrounding the circuitry and the wires is a shield which penetrates ~0.2 cm beyond the sphere. The shield is driven at the sphere potential over the full bandwidth, thus reducing any unwanted capacitance between the sphere surface and the electronics to less that 1 pF. The shield also reduces cross talk between spheres to a negligible level.

braid which is exposed to the plasma. The exposed conductor is segmented into several sections that have controlled potentials. Figure 4 is a detailed diagram of the wire boom sensors.

The electric field is derived from the difference in potential between two probes in voltage mode. The probe potential is determined by a balance of electron current, ion current, photo-electron emission, secondary electron emission, and a bias current (Mozer, 1973). The probe surfaces are coated with carbon that produces known photo-emission characteristics. Bias currents are adjusted from −100 nA to 100 nA in steps of ~0.8 nA to minimize errors. Bias tables for low and high plasma densities, eclipse and sun, and spacecraft configuration are stored on-board and are automatically adjusted as the spacecraft passes through the terminator.

The ‘stubs’ are 2.4 m sections of wire exposed to the plasma immediately beside the sphere. A stub section is added to the outside of the outer sphere for symmetry. The outer conductive surface of the stubs are driven at a fixed potential with respect to the sphere that is adjustable from −2.5 V to 2.5 V. By holding the stub potentials fixed with respect to the nearby sphere, the photo emission current between the sphere and the nearby wires can be controlled so that photo emission modulations on the wire minimally effect the potential of the sphere. The voltage control of the stubs is resistively coupled to maintain electrical stability.

The three 'guards' are 10-cm sections immediately beside the stubs. The potentials of the guards are adjustable from -10 V to 0 V with respect to the outer most probe potential. The guards are typically biased at -5 V to restrict photoemission current between the spacecraft and the spherical probes and between the two probes. Optimized stub and guard biases are also stored in the on-board bias tables.

The spin axis booms are rigid, 4-m stacers that have a single sphere with short (20-cm and 10-cm) stubs and no guards. The preamplifier design of the axial spheres is identical to the inner radial spheres. The axial probes can operate in voltage or current mode.

3.2. Preamplifiers

Plasma resistance to the spherical probes is expected to vary from $\sim 10^6\ \Omega$ to $>10^9\ \Omega$ and the capacitive coupling to the plasma is typically $\sim$5 pF. Preamplifiers located inside the spherical probes are designed to have very low stray capacitance (<1.0 pF) and very high input resistance ($>10^{11}\ \Omega$) so that electric fields from DC to $\sim$2 MHz can be measured. An aluminum shield (Figure 4) electrically shields the probe surface from the circuitry inside the probe and the wire which runs through the probe. The same cover also provides radiation shielding. Radial and axial preamplifier response is diagramed in Figure 5(a) and the signal processing coverage of the electric field is diagramed in Figure 5(b). Signals less than $\sim$300 Hz are typically resistively coupled to the plasma while those greater than $\sim$300 Hz are typically capacitively coupled to the plasma. The drop in preamplifier gain to 0.8 at frequencies greater than 300 Hz is due to the cross over from resistive to capacitive coupling.

In current mode, sensors 6, 7, 9, and 10 have a dynamic range from $\sim$0.5 nA to 2×10^4 nA representing a density range from $\sim$0.2 cm^{-3} to $\sim$104 cm^{-3} with typical auroral electron temperatures ($\sim$1 eV). Spheres 2 and 3 measure from $\sim$10 nA to 5×10^5 nA for low-altitude coverage. The bias voltages on sensors 2, 3, 6, and 7 can be set between 0 V and 20 V with respect to one of two base potentials: (1) the nearby voltage mode sensor or (2) the potential derived from 0.8 the nearby voltage sensor and 0.2 the spacecraft potential. Initial testing on orbit has shown the first base potential can cause undesirable spacecraft charging so the latter base potential is always used. Sensors 9 and 10 are always referenced from the payload and can be biased from -5 V to 45 V.

3.3. Boom Electronics Boards (BEBs)

A BEB is located in each of the four wire boom deployment units and one is located on the radiation shield of the main spacecraft to control the axial booms. The boom electronics boards perform the following functions: (1) receive commands from the

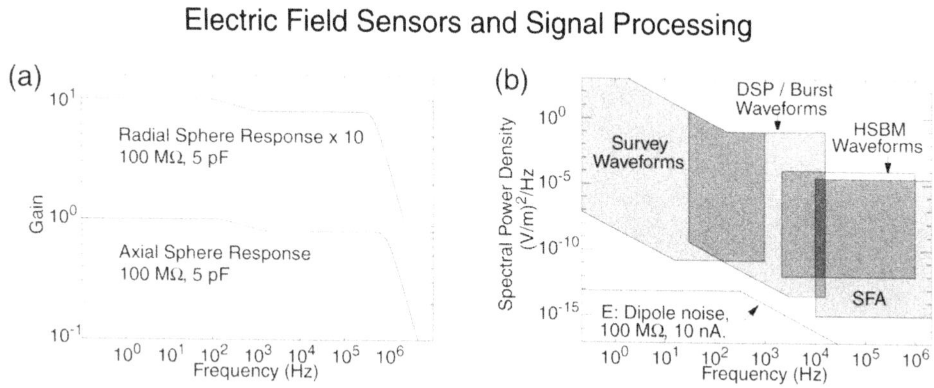

Figure 5. (a) The sensor preamplifier response as a function of frequency under 100 MΩ, 5 pF source impedance. The drop of gain from 1.0 to 0.8 at ~300 Hz represents the cross over from resistive to capacitive coupling. The response on radial booms has an effective a two-pole roll off at ~500 kHz due to losses driving the 28-m cable. The axial boom (2-m cable) preamplifier response has twice the bandwidth. (b) The theoretical noise level of the 56-m electric field antenna system and the dynamic range of survey waveform, DSP, SFA, and HSBM signal processing systems.

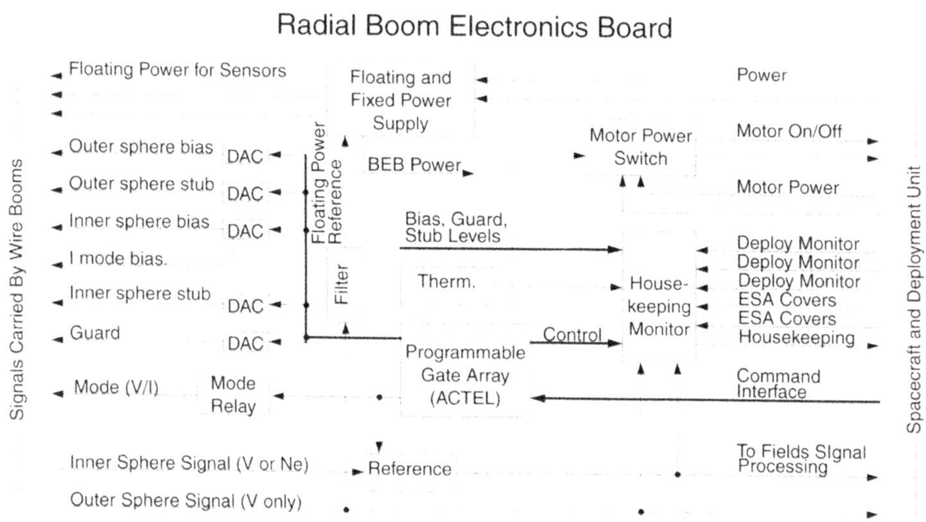

Figure 6. A block diagram of the radial boom electronics board.

instrument data processing unit (IDPU) including bias levels for the sensors, stubs, and guards, (2) provide analog bias voltages and currents to the sensors, (3) provide power to the sensor preamplifiers, (4) distribute the analog signals from the sensors to the fields signal processing system, (5) turn on and off wire boom motors from IDPU command, and (6) return housekeeping signals of bias levels, temperatures, and the state of deployment of the wire booms.

A block diagram of the radial BEB is in Figure 6. Commands from the IDPU are sent by a serial interface to a programmable gate array (all programmable gate arrays in FAST are Actel 1020 series) which contains all of the digital logic. Each BEB has a unique address so that it can be individually commanded. Commands include the bias, guard, and stub levels which are fed to an 8-bit digital to analog converter (DAC).

A floating power converter provides power for the BEB and the sensor preamplifers. The sensor power supply has a reference to the sensor potential with a range from -52 V to 52 V. The DC to $\sim$300 Hz dynamic range of the sensors is from ~ -45 V to 45 V which can measure ± 1.6 V m^{-1} on the 56-m dipoles and $+11$ V m^{-1} on the 8-m axial dipole. The 5-m dipoles saturate at the same electric field amplitudes as the 56-m dipoles, except for short wavelength emissions. The sensors have a operating range of ~ -8 V to 8 V for signals $>\sim$ 300 Hz. The 56-m and 5-m dipoles can measure up to 300 mV/m wave electric fields and the 7.7-m axial dipole can measure up to 2 V m^{-1} wave electric fields.

4. Signal Processing and Data Products

We start with a discussion of science modes (Section 4.1). The signal processing descriptions are organized following the diagram in Figure 3: (4.2) survey waveforms, (4.3) burst waveforms, (4.4) DSP: LF spectra, (4.5) SFA: HF spectra, (4.6) PWT, (4.7) HSBM: HF digital waveforms, (4.8) BBF: HF power, frequency, and phase, and (4.9) the wave-particle correlator. Several signal processing systems may share common resources. For example, the digital signal processor uses burst waveform data to produce the survey spectra.

4.1. Science Modes

The FAST instruments were defined under the concept of science 'modes', whereby the instruments can be configured to emphasize specific scientific investigations. Science modes also allow for optimizing investigations at various local times and altitudes as well as performing follow up investigations of any phenomena discovered during the mission. For example, a science mode that studies auroral kilometric radiation would emphasize the high-frequency signal processing. Another use of the modes is to control the overall data rate.

The Fields Instrument was designed with several layers of flexibility to accommodate a variety of science modes. The first layer of flexibility is in the sensor configuration. Six of the ten electric field sensors can be operated in one of two configurations. Next, a series of multiplexors allows selection of raw sensor signals, differential (electric field) signals, search coil magnetometer signals, and the fluxgate magnetometer signals into several signal processing systems. Finally, each of the signal processing systems has a variable data rate. The configuration of the sensors, the signal selection, and the individual data rates of each signal processing system are the main elements of a science mode.

4.2. Survey Waveforms

All of the survey waveform signals share a single 16-bit A/D converter (Crystal Semiconductor CS5016) that constantly samples at 32 768 samples s^{-1}. The signals are filtered with 4-pole Bessel filters (1 kHz or 250 Hz) that are in a hybrid package (~1 g, 7.2 mW each). The survey waveform conversion rate is configurable by factors of two from 16 samples s^{-1} to 2048 samples s^{-1} for the 1 kHz waveforms, and from 4 samples s^{-1} to 512 samples s^{-1} for 250 Hz waveforms. The signals are under-sampled except at the maximum rate.

Eighteen low-frequency analog signals are processed as survey waveforms:

- Three long-baseline DC electric field signals (1 kHz).
- Four configurable signals that are either short baseline DC electric field signals, sensor voltages, or Langmuir probe (plasma density) outputs (1 kHz).
- Three sensor voltages (250 Hz).
- Three DC fluxgate magnetometer signals (250 Hz).
- Three AC search coil signals (1 kHz).
- Two LF (~1 kHz–16 kHz) wave power signals (128 Hz).

The signal selection, sensitivities, dipole lengths, and the dynamic range of the survey signals are summarized in Table I. The signals that are most often selected in the science modes are shaded. Electric field measurement and Langmuir probe data are plotted in panels (a) and (b) of Figure 12.

The three-axis fluxgate magnetometer signals are converted at rates from 16 samples s^{-1} to 2048 samples s^{-1} before being input into a recursive digital filter that is in a programmable gate array. The recursive filter acts as a one-pole low-pass filter at $f_0 = -\ln(15/16) f_{\text{sample}}/2\pi$, which is ~21 Hz at maximum sample rate. The output rate of the fluxgate magnetometer signals are at $\frac{1}{4}$ the sample rate varying from 4 samples s^{-1} at the minimum rate to 512 samples s^{-1} at the maximum rate. Fluxgate magnetometer data are plotted in Figure 12(c).

4.3. Burst Waveforms

There are eight 16-bit (Crystal Semiconductor CS5016) A/D converters dedicated to burst waveforms which simultaneously sample at 32 768 samples s^{-1}. Up to 14 signals out of 40 signals (Table II) can be selected for A/D conversion. Signal selection is set in the science mode. Two of the eight A/D converters can be configured to operate in two different ways. Each can sample one signal at 32 768 samples s^{-1} (~16 kHz Nyquist) or four signals at 8196 samples s^{-1} (~4 kHz Nyquist). The burst waveforms can come in one of two configurations: (1) eight waveforms at 32 768 samples s^{-1} or (2) six waveforms at 32 768 samples s^{-1} and eight at 8196 samples s^{-1}. Wave-forms sampled at 32 768 samples s^{-1} are filtered by 6-pole, 16 kHz Butterworth filters in hybrid packages. Waveforms sampled at 8196 samples s^{-1} are filtered by a 4-pole, 4 kHz Bessel filter in hybrid packages.

Table II summarizes the available signals and their properties. The entries ending with HG are AC coupled, high-gain signals with a frequency band from ~3 kHz

TABLE I

Survey electric and magnetic field waveforms

Svy. pac-ket	1 kHZ waveform E	Dipole (m)	Sensitivity (mV m^{-1})	Range (V m^{-1})	Alt.: Plasma Density	Range (nA)	Alt.: Sensor Volt	Range (v)	Alt.: Wave-form	Di-pole (m)	Range (V m^{-1})
1 1	**V1–V4**	29									
2 1	**V5–V8**	56	0.05	± 1.6							
3 1	**V9–V10**	7.7	0.36	± 11	Ne9	$0.5-2\times10^4$	V9	± 45			
4 2	**V1–V2**	5	0.5	± 1.6	Ne2	$10-5\times10^5$	V2	± 45	V1-V3	28	± 1.6
5 2	**V3–V4**				Ne3		V3	± 45	V2-V4	23	± 1.6
6 2	V5–V6	5	0.5	± 1.6	**Ne6**	$0.5-2\times10^4$	V6	± 45	V5-V7	51	± 1.6
7 2	V7–V8	5	0.5	± 1.6	**Ne7**	$0.5-2\times10^4$	V7	± 45	V6-V8	51	± 1.6

Svy. Pac-ket	1 kHz Waveform Search	Search Coil Length	Sensitivity (nT-Hz)	Range (nT-Hz)
8 3	**Mag1ac**	7″	8.1×10^{-2}	$\pm 2.7\times10^3$
9 3	**Mag2ac**	7″	8.1×10^{-2}	$\pm 2.7\times10^3$
10 3	**Mag3ac**	21″	4.2×10^{-2}	$\pm 1.4\times10^3$

Svy. Pac-ket	205 Hz Signals	Freq. Range (kHz)	Range Minimum (mV m^{-1})	Range Maximum (mV m^{-1})	205 Hz Plasma Density	Range (nA)	250 Hz Sensor Volt	Range (v)
11 3							**V4**	± 45
12 3							**V8**	± 45
13 3					Ne10	$0.5-2\times10^4$	**V10**	± 45
14 3	**VLF1**	1-16	0.01	200				
15 3	**VLF2**	1-16	0.01	200				

Svy. Pac-ket	DC Waveform Fluxgate	Sensitivity (nT)	Range (G)
16 1	**Mag1dc**	2	± 0.6
17 1	**Mag2dc**	2	± 0.6
18 1	**Mag3dc**	2	± 0.6

to 16 kHz. The high-pass filter at 3 kHz has one pole. Otherwise, the signals are DC coupled. The sensitivity in Table II is the one-bit level, while the range is limited either by the maximum range of the sensor or the A/D converter. A/D converters 7 and 8 can be multiplexed to receive four 4 kHz bandwidth signals. If the signal is available at 4 kHz, an 'm', standing for 'multiplexed', is added (e.g., 8 m). Otherwise the signal is available at 16 kHz. A typical configuration is

TABLE II

	Signal E	Dipole (m)	Sensitivity (μV m^{-1})	Range* (V m^{-1})	A/D		Signal	Sensitivity (mV nA^{-1})	Range (V nA^{-1})	A/D band
1	**V1-V4HG**	29	4.7	± 0.2	1	21	V1	1.37	± 45	1
2	**V1-V2HG**	5	27.5	± 1.1	5	22	V2	1.37	± 45	5
3	V3-V4HG	–	–	–	6	23	V3	1.37	± 45	6
4	**V5-V8HG**	56	2.5	± 0.1	2	24	V4	1.37	± 45	3
5	V1-V4	29	95	±v1.6	1,8 m	25	V5	1.37	± 45	2
6	V5-V8*	56	49	± 1.6	2,8 m	26	V6	1.37	± 45	7
7	**V9-V10**	7.7	357	± 11	3,8	27	V7	1.37	± 45	8
8	V1-V2*	5	549	± 1.6	5,7 m,8 m	28	V8	1.37	± 45	4
9	V3-V4*	–	–	–	6,7 m.8 m	29	V9	1.37	± 45	4
10	V5-V6*	5	549	± 1.6	7,7 m,8 m	30	V10	1.37	± 45	8 m
11	V7-V8*	5	549	± 1.6	8,7 m,8 m	31	Ne2	10	5×10^5	5
12	V1-V3*	28	98	± 1.6	5,7 m,8 m	32	Ne3	–	–	6
139	V2-V4*	23	120	± 1.6	6,7 m.8 m	33	Ne6	0.5	2×10^4	7,7 m,8 m
14	V5-V7*	51	54	± 1.6	7,7 m,8 m	34	Ne7	0.5	2×10^4	8,7 m,8 m
15	V6-V8*	51	54	± 1.6	8,7 m,8 m	35	Ne9	0.5	2×10^4	4
16	VQUAD	–	–	–	7,8 m	36	Ne10	0.5	2×10^4	8 m

	Signal B	Search Coil	Sensitivity (nT-Hz)	Range (nT-Hz)	A/D		Signal	(See BBF)		A/D Band
17	**Mag1ac**	7″	8.7×10^{-2}	2850	7 m	37	BBF1			
18	**Mag2ac**	7″	8.7×10^{-2}	2850	7 m	38	BBF2			
19	**Mag3ac**	21″	4.2×10^{-2}	1376	4,6,7 m	39	BBF3			
20	**PWT**		See SFA	–	–	40	BBF4			

shaded. VQUAD is the measurement of the quadrupole signal (V1+V4−V5−V8). PWT, the plasma wave tracker, and the BBF channels described later.

The burst digital data are continuously available to the instrument data processing unit and to the DSP at an overall data rate of 4.194 Mbit s^{-1}. The periods of data capture are selected by the instrument data processor. The selection criteria are from trigger signals supplied by the Fields Instrument and the particle detectors.

4.4. Digital Signal Processor: LF Spectra

The primary function of the DSP is to provide continuous coverage of the spectral power density of the electric and magnetic field in the frequency range up to 16 kHz. It also has three other optional functions: provide spectral power density of the high-frequency (~1 kHz to 1 MHz) electric and magnetic field, perform cross-spectral analysis of the low-frequency and high-frequency electric field pairs,

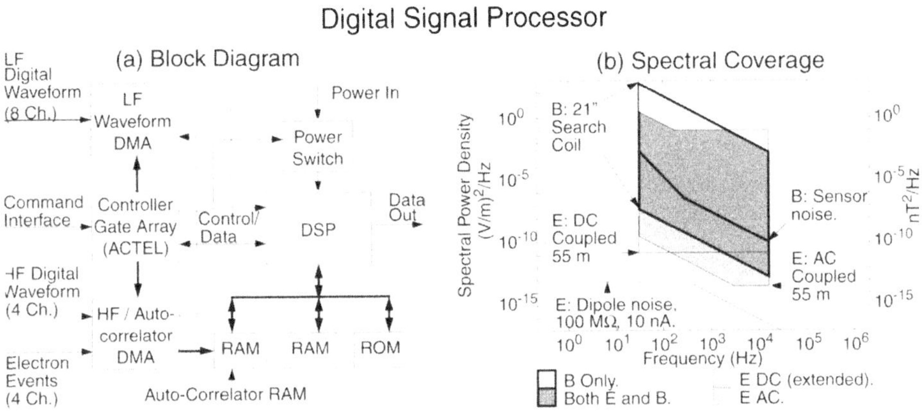

Figure 7. (a) A block diagram of the DSP. The DSP receives data from the burst A/D converters (low-frequency), the HSBM (high-frequency), and the electron electrostatic analyzers. The primary function of the DSP is to provide continuous 16 Hz–16 kHz wave power spectra in the auroral zone. (b) The dynamic range coverage versus frequency for DSP spectra. The DSP averages FFTs of electric and magnetic waveforms. The time resolution of the 1024 point FFTs varies from 31 ms to 4 s.

and calculate the auto-correlation function of four fixed energy channels from the electron spectrographs.

The LF spectrum analysis averages 2^n (n is configurable from 0 to 7), 1024-point Fast Fourier Transforms (FFT) of the digital waveforms from the burst A/D converters. The resulting spectra have 32 Hz resolution in frequency and from 32 ms to 4 s resolution in time. The selection, sensitivity, and range of the burst waveforms are described in Table II.

A block diagram of the DSP is in Figure 7(a). The eight burst waveforms are fed into a dedicated Direct Memory Access (DMA) designed into a programmable gate array. A second custom DMA was designed to access the high speed burst memory waveforms (1 MHz waveforms at 2 Msamples s^{-1}) and four channels of the electron spectrograph. All of the data passes from the DMAs into the RAM through single high-speed serial interface in the DSP. In all, there are sixteen inputs which are individually selectable.

A third programmable gate array acts as a controller. Its primary functions are to (1) direct start up, reprogramming, and resets of the DSP, (2) receive commands from the instrument data processing unit (IDPU), (3) provide timing and control of the DMAs, (4) detect single event latch up, single event upset, or malfunction of the DSP and restart, and (5) hardware protect the code area in RAM.

The DSP processor is a 32-bit floating point ATT-DSP32C with an input clock at 32 MHz (below the 50 MHz maximum). Radiation testing, performed at the University of California at Berkeley, determined that the total dose tolerance exceeds 200 kRad. Four radiation hardened static CMOS RAMs were used to make a 32-bit by 32 Kbit memory. Bi-polar ROMs are used to hold the code.

Due to the high power consumption (2.5 W) and slow access speed (100 ns) of the ROMs, the start-up sequence copies the ROM code into the RAM code area then turns off the ROMS. The code can be augmented or patched by the IDPU. The code area is hardware protected after the start-up sequence is completed.

The dynamic range of the DSP LF spectra are displayed in Figure 7(b). The data are logarithmically compressed to 8 bits. The electric field spectra range from 2×10^{-13} (V m^{-1})^2 Hz^{-1} to 2×10^{-3} (V m^{-1})^2 Hz^{-1} on the 56-m high gain (V5-V8HG) signal. The 5m DC-coupled signals have a dynamic range starting at 10^{-10} (V m^{-1})^2 Hz^{-1}. The range of the signal processing is optimized for auroral processes which are well above the sensor noise levels. The magnetic spectra of the 21″ search coil range from $\sim 1 \times 10^{-10}$ (nT)^2 Hz^{-1} to $\sim 5 \times 10^{-3}$ (nT)^2 Hz^{-1} at 1 kHz. The search coil dynamic range is limited by the sensor noise at frequencies greater than ~500 Hz. Auroral electric field spectra from the DSP are displayed in Figure 12(f).

4.5. Swept Frequency Analyzers: HF Spectra

All high-frequency signals are from the high-frequency analog unit (Figure 3) which is co-located with the SFA. The high-frequency analog unit has AC-coupled differential amplifiers which form electric field signals and analog switches for mode configuration. Table III describes selection and the sensitivities of the signals.

The primary functions of the SFA are (1) to provide continuous coverage of the high-frequency (~10 kHz–2 MHz) electric and magnetic fields as survey data, (2) provide high time resolution (62.5 ms) spectra of the electric and magnetic fields as burst data, and (3) provide fine frequency resolution observations of narrow-band emissions such as auroral kilometric radiation. The third function is designated as the Plasma Wave Tracker.

A block diagram of one SFA channel is given in Figure 8(a). The SFA has four channels. In three of the channels, the high-frequency signal is passed into a 3-pole low-pass filter at 2 MHz. The fourth channel, which can be used by the plasma wave tracker, has a 4 MHz filter. The filtered signal is mixed with a sweeping reference (10.7–12.7 MHz in normal operation), passed through a 10.7 MHz crystal filter with 15 kHz bandwidth, then mixed again at 10.65 MHz to produce a 50 kHz intermediate frequency. The resulting signal is rectified and logarithmically amplified before digital (8-bit) conversion by the high-frequency A/D converter (see Figure 2). The 256-point HF spectra have ~8 kHz steps (over-sampling the 15 kHz band width) covering from 0 to 2 MHz. Since the sweeping reference is produce by a digital frequency synthesizer, the sweep can be configured to have maximum range of 500 Hz, 1 MHz, 2 MHz, or 4 MHz. The sweep rate can be configured to 31.25 ms or ~62.5 ms (typical operation).

The survey data are an average 2^n sweeps, where n is set from 0 to 7. The sweep rate changes from 'slow survey' to 'fast survey', usually from $n = 6$ to $n = 3$, thus increasing the data rate and time resolution of the sweeps (most often

TABLE III

HF signal processing

Ch.	Signal SFA	Dipole (m)	Sensitivity $(V\ m^{-1})^2\ Hz^{-1}$ $(nT^2\ Hz^{-1})$	Range $(V\ m^{-1})^2\ Hz^{-1}$ $(nT^2\ Hz^{-1})$	Frequency max. (MHz)	Alternative signal	Dipole (m)	Sensitivity $(V\ m^{-1})^2\ Hz^{-1}$	Range $(V\ m^{-1})^2\ Hz^{-1}$	Frequency max. (MHz)
1	V1-V4	29	4.0×10^{-15}	6.0×10^{-7}	2	V1-V2	5	2.0×10^{-13}	2.0×10^{-5}	2
2	V5-V8	56	1.0×10^{-15}	1.6×10^{-7}	2	V7-V8	5	2.0×10^{-13}	2.0×10^{-5}	2
3	Mag3ac	21″ SC	1.0×10^{-11}	1.0×10^{-4}	0.6	V5-V6	5	2.0×10^{-13}	2.0×10^{-5}	2
4	V9-V10	7.7	3.0×10^{-13}	3.0×10^{-5}	4	V3-V4	–	–	–	–
4						V1-V4	29	4.0×10^{-15}	6.0×10^{-7}	2
4						V1-V2	5	2.0×10^{-13}	2.0×10^{-5}	2

Ch.	Signal HSBM	Dipole (m)	Sensitivity $(V\ m^{-1})$ or (nT)	Range* $(V\ m^{-1})$ or (nT)	Frequency max. (MHz)	Alternative signal	Dipole (m)	Sensitivity $(V\ m^{-1})$	Range* $(V\ m^{-1})$	Frequency max. (MHz)
1	V1-V4	29	0.4	220	2	V1-V2	5	2.0	500	2
2	V5-V8	56	0.2	110	2	V7-V8	5	2.0	500	2
3	Mag3ac	21″ SC	4.0×10^{-3}	2.0	0.6	V5-V6	5	2.0	500	2
4	V9-V10	7.7	1.25	750	4	V3-V4	–	–	–	–

Ch.	Signal HF Power	Dipole (m)	Sensitivity $(V\ m^{-1})$ or (nT)	Range* $(V\ m^{-1})$ or (nT)	Frequency max. (MHz)	Alternative signal	Dipole (m)	Sensitivity $(V\ m^{-1})$	Range* $(V\ m^{-1})$	Frequency max. (MHz)
1	V1-V4	29	0.2	220	2	V1-V2	5	1.0	500	2
2	V5-V8	56	0.1	110	2	V7-V8	5	1.0	500	2
3	Mag3ac	21″ SC	2.0×10^{-3}	2.0	0.6	V5-V6	5	1.0	500	2
4	V9-V10	7.7	0.75	2000	4	V3-V4	–	–	–	–

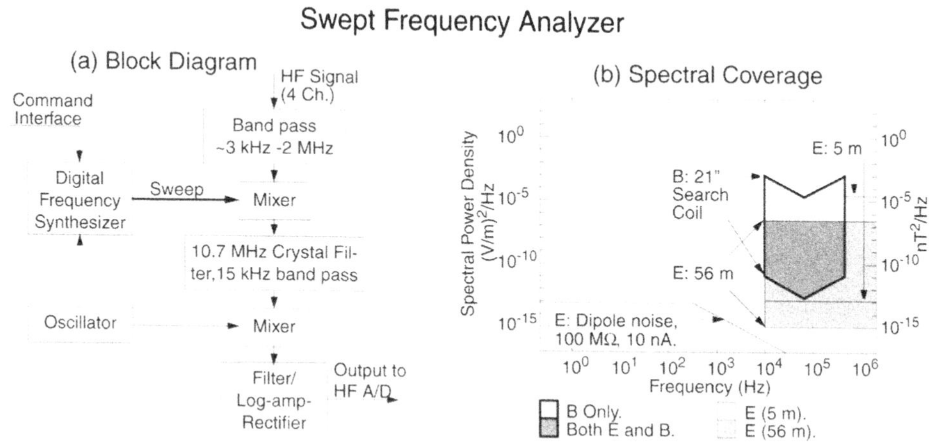

Figure 8. (a) A block diagram of the SFA. Three of the four channels are filtered to 2 MHz and one channel to 4 MHz. The latter channel can be used by the plasma wave tracker by setting the digital frequency synthesizer to a fixed frequency. (b) The spectral coverage and dynamic range of the SFA. The electric field range is from 10^{-15} (V m^{-1})^2 Hz^{-1} (56-m dipole) to 10^{-5} (V m^{-1})^2 Hz^{-1} (5-m dipole). The 21″ search coil has a peak sensitivity at ~65 kHz with a frequency range to ~600 kHz.

4 s in slow survey and 0.5 s in fast survey). Burst data transmits every sweep. The spectral coverage of the SFA is displayed in Figure 8(b). The usual gain setting in the auroral zone has a dynamic range from 10^{-15} to 10^{-7} (V m^{-1})^2 Hz^{-1} for electric fields and from 10^{-12} to 10^{-4} nT^2 Hz^{-1} (at 100 kHz) for magnetic fields. SFA data are displayed in Figure 12(e).

4.6. Plasma Wave Tracker

The primary function of the PWT is to provide fine-frequency resolution spectra of narrow- band emissions. This is accomplished using the fourth channel of the SFA. The signal from the digital frequency synthesizer is set at a fixed frequency (f_0) that is (1) fixed by configuration, (2) dynamically set at the electron cyclotron frequency, or (3) dynamically set by the number of zero crossings in the wave form. The fixed frequency is the center of the 15 kHz band.

After passing through the 10 700 MHz crystal filter (Figure 8(a)), the signal passes to the second mixer which is set at 10 692 MHz when configured for the plasma wave tracker. The resulting signal (~0.5 kHz to 15.5 kHz) represents the frequency band $f_0 - 7.5$ kHz to $f_0 + 7.5$ kHz. The PWT signal then is digitized as a burst waveform (see Table II, entry 20) and/or processed by the DSP. The frequency resolution of the PWT signal is limited by the jitter in the digital frequency synthesizer to +50 Hz. An example of PWT data is given in Figure 12(d).

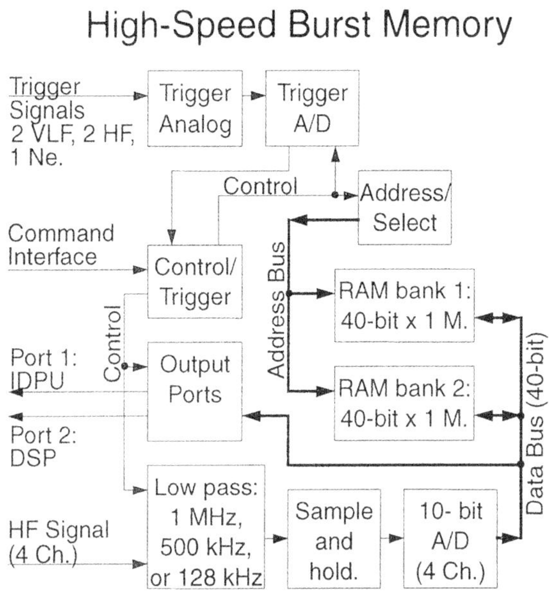

Figure 9. A block diagram of the HSBM. Four high-frequency channels are low-pass filtered then digitized. The data are selected by high- or low-frequency wave event or plasma density cavity by a dedicated trigger system and are stored in a 10 Mbyte RAM. The RAM buffers can be output as waveform data (Port 1) or transferred to the DSP which produces a average of FFTs.

4.7. High Speed Burst Memory: HF Digital Waveforms

The HSBM produces 0.5 μs resolution digital waveforms of four signals, typically three electric field and one magnetic field, with $\sim$0.1% coverage in the auroral zone ($\sim$2 s out of 30 min). Alternatively, the HSBM can supply the DSP (described above) with digital HF waveforms to be converted into HF spectra. The HSBM is located in the Fields Instrument signal processing and is not part of the main instrument burst memory. A block diagram is in Figure 9. Signal selection and dynamic range are in Table III.

The HSBM has three main sub-systems. The analog section has selectable three-pole low-pass filters of 125 kHz, 500 kHz, and 1 MHz corresponding to three selectable A/D speeds of 250 ksample s^{-1}, 1 Msample s^{-1}, or 2 Msample s^{-1}. Four 10-bit MP7694 A/D converters are augmented with a high-speed sample and hold. The four A/D converters are on a 40-bit bus that is continuously written to a 10 Mbyte RAM. The trigger system uses the rectified and logarithmically amplified low-frequency wave power (see survey waveforms), the high-frequency wave power (see later), and the plasma density (see survey waveforms) as inputs. The trigger A/D converter alternately samples two of the selected inputs labeled 'trigger A' and 'trigger B', often set to one low-frequency and one high-frequency power level.

The digital logic is contained in three programmable gate arrays and works as follows. The RAM is divided into four buffers, one of which continuously accepts data (input buffer), two are holding buffers with the highest trigger A and trigger B levels, and the forth buffer outputs data (output buffer). If the trigger level exceeds the level in one of the holding buffers, the instrument (1) waits for $\frac{3}{4}$ of the input buffer to be written keeping $\frac{1}{4}$ of the buffer for data prior to the trigger event and (2) swaps the input buffer and a holding buffer thus overwriting the holding buffer. This process continues until the output buffer has been completely read. At that point, one of the holding buffers, alternating between the two, becomes the output buffer. The new holding buffer is assigned a zero trigger level.

The logical process above has several additional features. One can assign a minimum trigger level to the HSBM so that a trigger event cannot occur unless the minimum requirement is met. There are two settings of trigger position within the buffer keeping either $\frac{1}{4}$ or $\frac{1}{2}$ of the buffer for data prior to the trigger. The triggers can be set 'retriggerable' or 'absolute'. The former allows for a trigger event to restart if the trigger level increases later in the event so that the peak is always at the trigger position. The latter does not allow for a restart. The maximum buffer size is 2.5 Mbytes. The buffer sizes can be $1/2^n$ of the maximum with n between 0 and 7. Finally, on can go into 'time-based' triggering whereby the triggers events occur evenly in time rather than by plasma wave amplitude. The last feature is used when the DSP generating spectra.

4.8. Broad-Band Filters: HF Power, Frequency, and Phase

The main function of the Broad-Band Filter (BBF) signal processing is to provide high-time resolution power, frequency, and phase information for high-frequency signals (Figure 10). Four AC-coupled, high-frequency signals are processed. The selection, sensitivities, and ranges are described in Table III. The selected signals are high-pass filtered at 200 kHz with a three-pole passive LCR filter. Since the sensor preamplifier response rolls off at $\sim$500 kHz, the BBF emphasizes auroral kilometric radiation which falls in the 200 kHz to 500 kHz band. The four signals are rectified and pseudo-logarithmically amplified and then passed to the high-frequency A/D converter (Figure 3) at typically 1 ms resolution. The BBF has a 60 dB dynamic range.

The selected analog signals are also fed to a comparator producing a digital representation of the zero crossing. The digital signals are input to counters which yield the frequency of a dominating, narrow band signal. Six pairs of the digital signals are processed by a phase discriminator which measures the relative phase of the signals to determine wave polarization. The zero-crossing counters and the phase measurements have 4 ms time resolution.

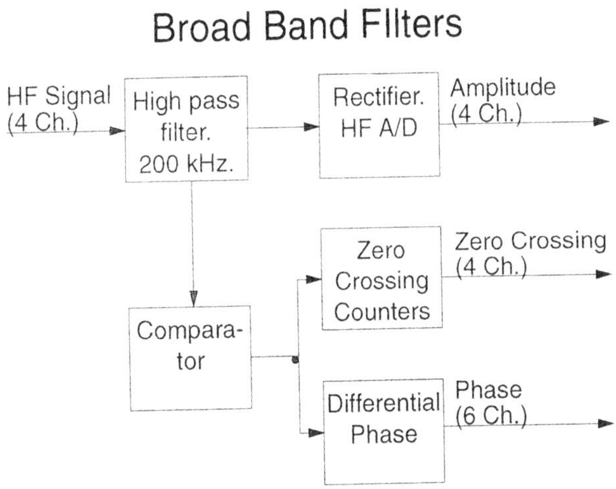

Figure 10. A block diagram of the broad band filter, phase, and zero crossing processing. Four high-frequency channels are high-pass filtered then rectified to measure the high-frequency wave amplitude at high time resolution. The high-frequency signals are digitized by a comparator.

4.9. Wave-Particle Correlator

The main scientific objective of the wave-particle correlator instrument is to identify the energy and pitch angle of particles that are interacting with waves. Measurement of the amplitude of particle oscillations and their phase relation with the wave allows for the evolution of the wave and distribution function to be studied in detail. Wave-particle correlator instruments provide a direct observation of wave-particle interactions. The wave-particle correlator has been described in a previous article (Ergun *et al.*, 1998), so we do not provide a full description here.

The wave-particle correlator detects oscillations in particle flux by integrating the product of the wave electric field and particle flux (Ergun *et al.*, 1991a; Lin *et al.*, 1995). This technique can be used to determine both the amplitude and phase (with respect to the wave phase) of oscillatory currents. 'Resistive' currents are in phase with the electric field and determine energy flow between the wave and particles (Ergun *et al.*, 1991b), whereas 'reactive' currents are in phase with wave potential and can indicate nonlinear kinetic processes such as particle trapping (Muschietti *et al.*, 1994).

One channel of the wave-particle correlator is diagramed in Figure 11. The electric field signal from the antenna is filtered to the desired frequency band. The instrument has selectable pass bands of 200 kHz to 2 MHz for Langmuir wave and auroral kilometric radiation studies, and 500 Hz to 20 kHz for lower frequency wave studies. The filtered electric field signal is fed into an analog phase splitter which consists of two all-pass filters that have phase responses which differ by 90° over a broad frequency range. (An all-pass filter is a unity gain circuit that has a non-zero phase response.) The digital output of the comparator represents the

TABLE IV

FAST fields instrument: mass and power

Component	Sub Component	Num ber	Mass	Power	Duty cycle	Orbit averaged power
	Radial boom/sensors	4	3.1 kg			
	Radial BEB	4	0.5 kg	0.5 W		
Electric Field	Axial boom/sensors	2	1.6 kg			
BEB/sensors	Axial BEB	1	0.5 kg	0.5 W		
	Totals (regulated power)			2.5 W		
	Total (unregulated power)		18.1 kg	2.8 W	0.25	0.70 W
	Senor		0.87 kg	0.45 W		
Search coil	Boom		1.87 kg			
	Totals (unregulated power)		2.65 kg	0.60 W	0.25	0.15 W
	Sensor		0.63 kg			
Fluxgate	Boom		1.78 kg			
(not including	Dirver electronics		0.90 kg	1.51 W		
ACS sensor.)	Totals (unregulated power)		3.31 kg	2.01 W	1.00	2.01 W
Signal	LF analog and A/D converters		1.76 kg	2.22 W		
processing	HF analog and SFA		0.88 kg	1.92 W	1.00 x	
	Digital signal proc.		0.40 kg	1.28 W	0.35 W	
	High speed burst memory		0.55 kg	0.65 W	and	
	Wave-particle correlator		0.80 kg	1.75 W	0.25 x	
	Totals (regulated power)			7.82 W	10.1 W	
	Totals (unregulated power)		4.39 kg	10.43 W		2.88 W
Totals			28.45 kg	15.9 W		5.74 W

polarity of the wave signal. The comparator output is then digitally integrated with electron events to make the correlation function.

5. Fields Instrument Performance

The Fields Instrument has met or exceeded all of the design specifications as outlined above, with the exception of the boom deployment. The radial wires holding sensors 3 and 4 jammed during initial deployment; the cause of the jam is unknown. Sensor 4 is exposed to the plasma near the spacecraft while sensor 3 is undeployed and is fixed at spacecraft ground. The impacts were minimal because of the flexible design of the Field Signal processing. They are:

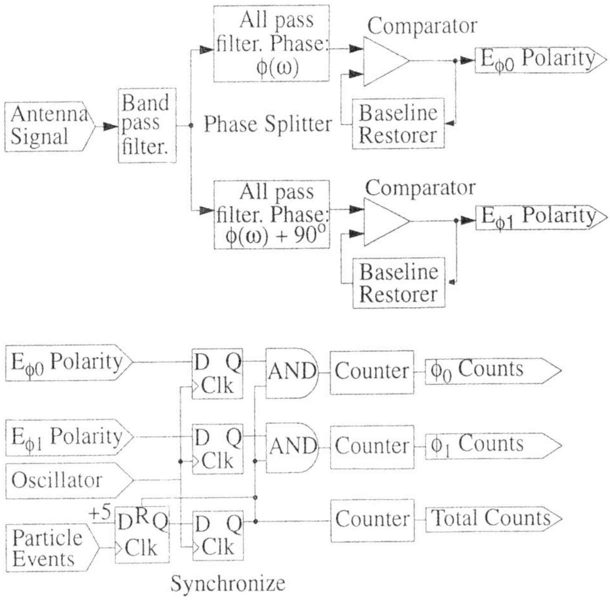

Figure 11. A simplified block diagram of a one channel wave-particle correlator. The electric field signal processing is diagrammed on top. Comparators produce two digital signals that represent polarities of the wave signal that has been split into two phases that differ by 90°. The wave polarity signals are correlated with the particle pulses (bottom) with digital logic. The ϕ_0*Count* (or ϕ_1*Count*) and is incremented only if the polarity of the ϕ_0 (or ϕ_1) wave signal is positive. The correlation measurements are (2ϕ_0*Count - Total Count*) and (2ϕ_1*Count - Total Count*).

- The DC (<300 Hz) contamination on sensor 4 was too high for scientific use. The spin-plane electric fields are calculated from linear combinations of signals V1, V5, and V8 (from Table I, signals 1, 2, and 5). There was little loss of accuracy and excellent scientific return (see Carlson *et al.*, 1998b, and references therein).
- The AC contamination on sensor 4 was remarkably low. Comparisons between V5-V8 and V1-V4 AC-coupled waveforms (burst and HSBM) and spectra (DSP and SFA) showed little or no increase in noise.
- Cross-correlation measurements are made between V5-V6 and V7-V8 which are not co-aligned. There was a loss of sensitivity in cross-spectral analysis.
- The sensor 10 axial boom deployment was delayed to avoid an a spin-axis tip of ~2.2° during the early phase of the FAST mission.

6. Mass and Power

Table IV summarizes the mass and power for the Fields Instrument. The radial boom sensors include the deployment mechanism, two sensors, the wire booms, and support structure for the radiation shield. The axial boom and sensor mass does

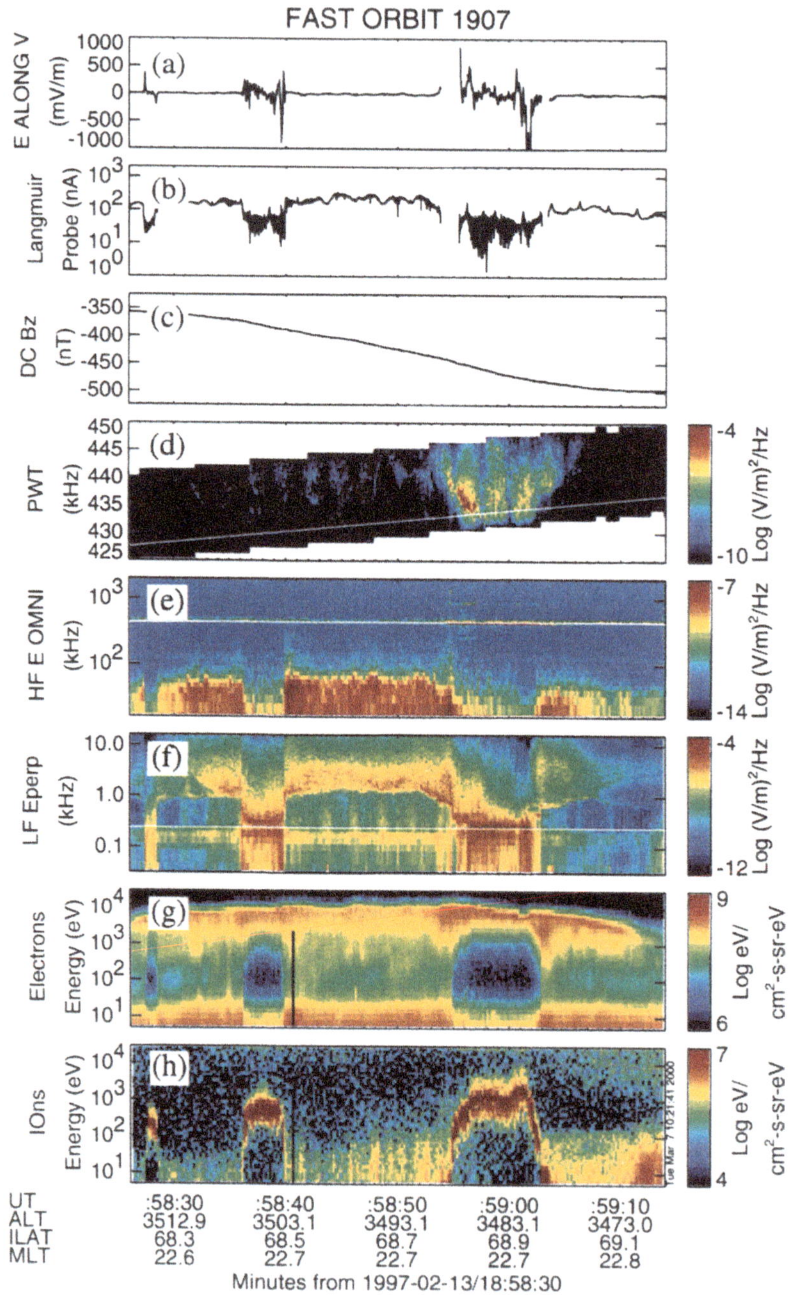

Figure 12. Data from the FAST satellite. (a) The DC electric field perpendicular to B projected along the velocity vector of the spacecraft. (b) The current to the Langmuir probe reflecting plasma density. Plasma density scales at ~ 1 cm^{-3} to 1 nA. (c) The magnetic field along the pay-load spin axis (nearly West). (d) Fine-frequency resolution data from the Plasma Wave Tracker. The electron cyclotron frequency is also plotted. (e) Electric field power spectra from the Swept Frequency Analyzer. (f) Low-frequency emissions (DSP data). The H^+ cyclotron frequency is also plotted. (g) Electron and (h) ion data from the energetic particle detectors.

not include the transmitter support tubes. The majority of the signal processing and electric field sensors are nominally on for ~25% of the orbit resulting in low orbit-averaged power consumption.

References

Carlson, C. W. *et al.*: 1998a, in R. R. Pfaff, J. E. Borovsky, and D. T. Young (eds.), 'Design and Applications of Imaging Plasma Instruments', *Measurement Techniques in Space Plasma*, AGU, Geophysical Monograph 102, p. 125.

Carlson, C. W., Pfaff, R. F., and Watzin, J. G.: 1998b, 'The Fast Auroral Snapshot Mission', *Geophys. Res. Lett.* **25**, 2013.

Ergun, R. E., Carlson, C. W., McFadden, J. P., Clemmons, J. H., and Boehm, L. H.: 1991a, 'Langmuir Wave Growth and Electron Bunching: Results From a Wave-Particle Correlator', *J. Geophys. Res.* **96**, 225.

Ergun, R. E., Carlson, C. W., McFadden, J. P., TonThat, D. M., Clemmons, J. H., and Boehm, M. H.: 1991b, 'Observation of Electron Bunching During Landau Growth and Damping', *J. Geophys. Res.* **96**, 11371.

Ergun, R. E., McFadden, J. P., and Carlson, C. W.: 1998, in R. F. Pfaff, J. E. Borovsky, and D. T. Young (eds.), 'Wave-Particle correlator Instrument Design', *Measurement Techniques in Space Plasmas* AGU Geophysical Monograph 102, p. 325.

Gurnett, D. A.: 1974, 'The Earth as a Radio Source: Terrestrial Kilometric Radiation', *J. Geophys. Res.* **79**, 4227.

Gurnett, D. A., Pfeiffer, G. W., Anderson, R. R. Mosier, S. R., and Cauffman, D. P.: 1969, 'Initial Observations of VLF Electric and Magnetic Fields with the Injun 5 Satellite', *J. Geophys. Res.* **74**, 4631.

Klumpar, D. M.: 1986, in T. Chang (ed.), 'A Digest and Comprehensive Bibliography on Transverse Auroral Ion Acceleration', *Ion Acceleration in the Magnetosphere and Ionosphere*, American Geophysical Union Monograph, p. 389.

Lin, R. P., Anderson, K. A., Ashford, S., Carlson, C., Curtis, D., Ergun, R., Larson, D., McFadden, J., McCarthy, M., Parks, G. K., Reme, H., Bosqued, J. M., Coutelier, J., Cotin, F., D'uston, C., Wenzel, K.-P., Sanderson, T. R., Henrion, J., Ronnet, J. C., and Paschmann, G.: 1995, 'A Three-Dimensional Plasma and Energetic Particle Investigation for the Wind Spacecraft', *Space Sci. Rev.* **71**, 125

McFadden, J. P., Carlson, C. W., and Boehm, M. H.: 1986, 'Field-Aligned Electron Precipitation at the Edge of an Arc', *J. Geophys. Res.* **91**, 1723.

Mozer, F. S.: 1973, 'Analysis of Techniques for Measuring DC and AC Electric Fields in the Magnetosphere', *Space Sci. Rev.* **14**, 272.

Mozer, F. S., Carlson, C. W., Hudson, M. K., Torbert, R. B., Parady, B., Yatteau, J., and Kelley, M. C.: 1977, 'Observations of Paired Electrostatic Shocks in the Polar Magnetosphere', *Phys. Rev. Lett.* **38**, 292.

Temerin, M. A., Carlson, C. W., Cattell, C. A., Ergun, R. E., McFadden, J. P., Mozer, F. S., Klumpar, D. M., Peterson, W. K., Shelley, E. G., and Elphic, R. C.: 1990, in T. Chang *et al.*, (eds.), 'Wave-Particle Interactions on the FAST Satellite', *Physics of Space Plasmas (1989)*, p. 343, Scientific Publishers, Inc., Cambridge, MA.

Temerin, M., Cerny, K., Lotko, W., and Mozer, F. S.: 1982, 'Observations of Double Layers and Solitary Waves in the Auroral Plasma', *Phys. Rev. Lett.* **48**, 1175.

DEPLOYMENT MECHANISMS ON THE FAST SATELLITE: MAGNETOMETER, RADIAL WIRE, AND AXIAL BOOMS

DAVID PANKOW
Samuel Silver Space Sciences Laboratory, University of California at Berkeley, U.S.A

ROBERT BESUNER
Besuner Consulting Services, Madera, California, U.S.A.

ROBERT WILKES
Lockheed Martin Space Mission Systems & Services, Houston, Texas, U.S.A.

ROBERT ULLRICH
Samuel Silver Space Sciences Laboratory, University of California at Berkeley, U.S.A.

Abstract. The Fast Auroral SnapshoT (FAST) satellite was launched by a Pegasus XL on August 21, 1996. This was the second launch in the NASA SMall EXplorer (SMEX) program. Early in the mission planning the decision was made to have the University of California at Berkeley Space Sciences Laboratory (UCB-SSL) mechanical engineering staff provide all of the spacecraft appendages, in order to meet the short development schedule, and to insure compatibility. This paper describes the design development, testing and on-orbit deployment of these boom systems: the 2 m carbon fiber magnetometer booms, the 58 m tip to tip spin-plane wire booms, and the 7 m dipole axial stiff booms.

Table of Contents

Space Science Reviews **98:** 93–111, 2001.

1. Introduction

1.1. Mission Background

The scientific objective of this magnetospheric physics mission was a detailed investigation of the Aurora Borealis, or 'Northern Lights'. The fields experiments (electric and magnetic) were constructed by the University of California at Berkeley (UCB), and Los Angeles (UCLA) respectively. The particles instruments were constructed by UCB and the University of New Hampshire in collaboration with Lockheed Palo Alto Research Laboratory. The instrument data processing unit was provided by UCB. The spacecraft bus, telemetry, and launch services were provided by the NASA Goddard Space Flight Center SMEX office. The science principal investigator is Dr C. W. Carlson of UCB, and the program is managed by the SMEX office.

The UCB design philosophy emphasizes the demonstration of design margins set by peer review. As a result, each boom system was extensively tested at a prototype level before the flight units were manufactured. Additionally, the design, assembly and testing of each boom mechanism was conducted by a single engineer solely responsible for its success.

1.2. Electric Field Sensor Description

Three orthogonal dipoles with six tip mounted sensors are needed for a vector measurement of the DC and AC electric fields in the plasma. Sounding rocket and early satellite experiments used stiff, deployable booms for the dipoles. Solar-thermal bending and vehicle dynamics severely limited these stiff booms to lengths of several meters, far short of the lengths desired for more precise physical measurements. In the evolution of these instruments, the preferred practical configuration has been found to be a spinning vehicle with four limp wires in the 'spin plane' and two stiff axial booms along the spin axis. The limp wires can be precisely positioned by centripetal acceleration, and are immune to the bending and buckling concerns in stiff booms. These lightweight wires allowed about a tenfold increase in the practical radial boom dipole lengths.

1.3. Spacecraft Dynamics

In practice, boom lengths are determined by the need for a spin stable vehicle. Briefly stated, a spinning body will be stable about the principal axis having the largest moment of inertia (or second mass moment), based on conservation of angular momentum and body-flexing dissipation of energy by the limp wire to a rotational energy minimum (Meirovitch and Calico, 1972). The spin stability ratio (which must be >1) is defined as the ratio of the moment of inertia about the spin axis to the larger of the two transverse axes ($I_S/I_{T_{max}}$), while the stability margin is defined as this ratio minus one. This means the radial wire booms improve stability

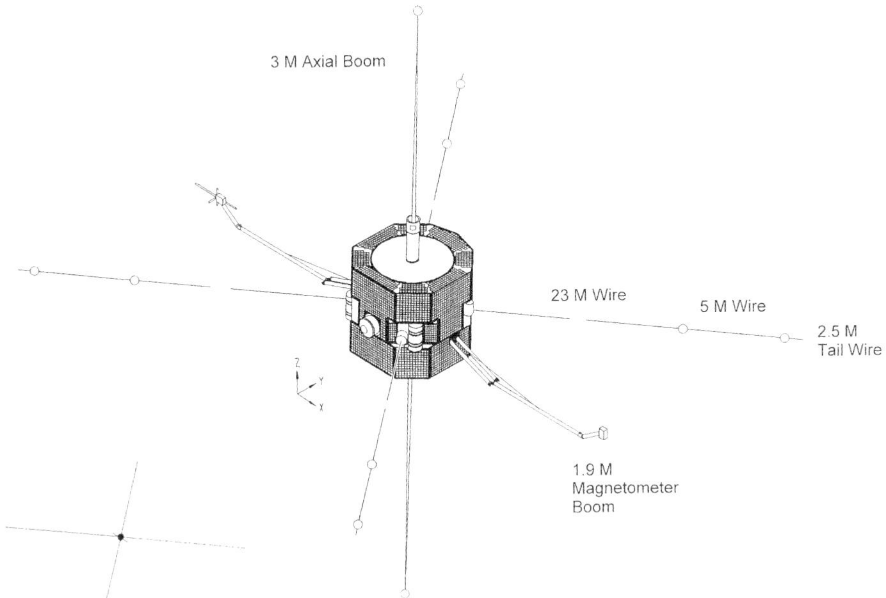

Figure 1. FAST Boom Geometry.

and can be quite long, while the axial booms are length limited because they reduce the stability margin by increasing the transverse moment of inertia.

The wire boom cables are essentially limp: any transient motions or oscillations are induced by spacecraft maneuvers. This pendulum behavior is dependent on the wire root or hinge attachment radius, the distance from the spin axis to the wire attachment (or exit point). On FAST, the radial boom stiffness was enhanced by providing trap doors at the vehicle skin, which increased this hinge radius. There is a practical (spacecraft attitude control system (ACS) stability based) bound on wire length, which is the geometry in which the spin plane torsional resonance equals the spin rate. This mode may be envisioned as a spin plane 'flutter' with all wires swinging in phase and a 180° phase shift for the center body (Lai and Bhavnani, 1975). The FAST boom geometry approached this limit, with a wire boom second mass moment that is 24 times that of the vehicle. The deployed wire boom plane was located close to the spacecraft Z axis center of mass to avoid spin axis tilt caused by wire boom mass moment asymmetries.

The axial booms must be sufficiently rigid to avoid elastic instability and subsequent collapse. As previously stated, the vehicle stability margin limits the boom length. In the mission planning stages, it was decided to include the stabilizing effect of wire booms in the overall moment of inertia calculations, to maximize the allowed axial boom length. In practice this increased the boom length from 2.6 m to 3.8 m each, which is a very significant improvement for minimizing the effects of vehicle photo-electron emission. Conventional wisdom would suggest that boom length might be increased by decreasing the boom mass, which will also decrease the stiffness. However, spin induced boom flexing amplifies the 'effective' boom second mass moment (Meirovitch and Calico, 1972). A boom cantilever resonance of four times the spin (as compared to a customary requirement of two) was selected to maximize boom length.

A systems issue is evaluation of the spin axis alignment budget. A list of many uncertainties, ranging from deployed boom straightness to alignment of the vehicle balance fixtures, will affect the alignment of the spin axis with the vehicle geometric axis. Simple addition of this list is far too conservative, and not warranted. If each of the uncertainties is assumed to have random clocking with respect to the spin axis, the resulting imbalance is one half the root square summation (RSS) of these residual inertia products. The traditional minimum requirement for the vehicle stability margin is 4%, based mostly in the uncertainties of mass moment measurements. Sensitivity of the spin axis alignment indicated that a more practical stability minimum was 8–10% for this satellite.

1.4. Boom Deployment Sequence

The deployment sequence for FAST released the magnetometer booms, the radial wire booms, then the axial booms. For both enhanced reliability and simplicity, these boom mechanisms are purposely designed without a retraction capability. The boom systems are manually rewound and reset after ground testing, and on-orbit retraction is neither possible nor necessary.

2. The Magnetometer Mechanisms

2.1. The Magnetometer Boom Configuration

System constraints imposed on the magnetometer booms were straightforward. They must be stiff compared to the other booms (to avoid dynamic interactions) and thermally stable to maintain alignment of the sensors. A two meter separation from the spacecraft was determined to be adequate to avoid the false signals related to spacecraft current loops and digital circuitry. Perhaps the greatest challenge in the design of these booms was fitting within the 50 mm irregular annulus between the solar array and the launch vehicle fairing.

Fluxgate, search coil, and attitude control system (ACS) magnetometers are installed on two deployable, articulated, carbon-fiber booms, approximately two meters in length. Significant features of the magnetometer booms include latchless hinges, low mass, and an innovative carbon-fiber/epoxy fabrication technique.

Each magnetometer boom assembly consists of a base assembly, dual inner boom segments, an outer boom segment, and saddles at the upper and lower corners of the spacecraft solar arrays. The base assembly includes a centrifugal brake to control deployment velocities, a restraint/deployment assist system, electrical

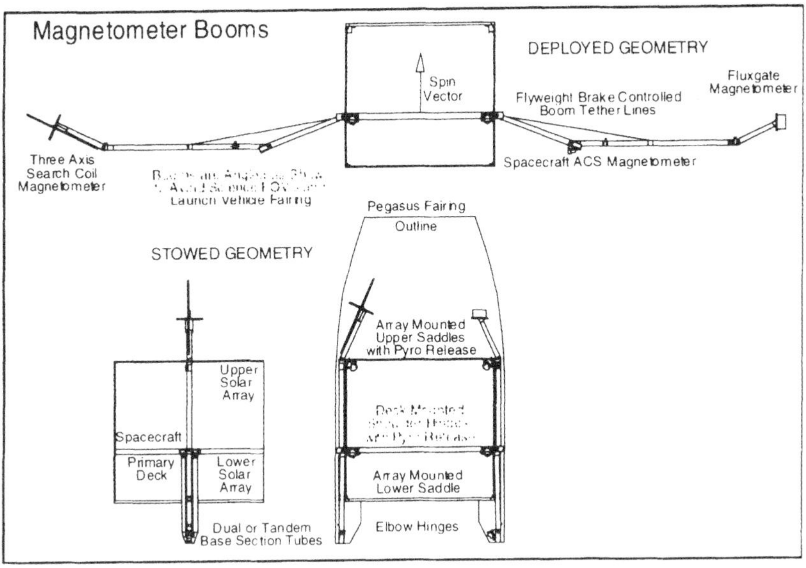

Figure 2. Magnetometer Boom Configuration.

connectors for spacecraft 'housekeeping' and magnetometer data, and the 'shoulder hinge'. The inner boom segment is composed of two carbon-fiber tubes, side by side, with space between them in which the outer boom segment nests. The outer boom segment is hinged to the inner segment and is comprised of carbon-fiber tubes with aluminum 'hard points' where the boom is restrained for launch dynamics. The outer boom elements are angled to clear the launch fairing of the Pegasus XL. The science magnetometers (fluxgate and search coils) are mounted at the ends of the booms. The ACS magnetometer is located on the elbow hinge of the fluxgate magnetometer boom.

The booms are mounted to the deck on opposite sides of the spacecraft. During launch, the booms were stowed against saddles on the sides of the spacecraft solar arrays, restrained by stainless steel cables at the spacecraft deck and by beryllium-copper rods at the upper array corners. The booms were preloaded against the deck and solar array edges to reduce the launch induced vibration amplitudes. This three-point bending geometry bowed the middle of the booms toward the spacecraft with a force at the deck of ~450 N. High force, short travel spring based deployment assist plungers were incorporated in the deck mounts, to overcome any binding forces that might develop in the mounts. These plungers also freed the centrifugal brakes, which were locked by the plungers for launch.

2.2. Magnetometer Boom Deployment

Once the spacecraft was in orbit and spinning at 20 rpm, the magnetometer booms were deployed by firing pyrotechnic cable cutters at the upper saddles and shoulder hinges. Centrifugal force motivated the deployment. The speed of deployment was controlled by a tether attached to the outer boom segment which spun a centrifugal (flyweight) brake in the base hinge assembly. The function of this brake is to provide a force balanced boom deployment environment, so that the booms could be deployed over a wide range of spin rates. Analysis and test set this range to be 12 to 60 rpm. The final deployed position of the booms was determined by hard stops in the hinges. These hard stops consist of v-blocks into which the round boom nests. There is no latching mechanism. The position along the spin (Z) axis of the line of force due to centripetal acceleration of each boom / magnetometer assembly is between the Z positions of the shoulder and elbow hinges. This causes the shoulder hinge to try to rotate in the spacecraft up direction and the elbow hinge to try to rotate in the spacecraft down direction, into their respective V-block stops. Torsion springs in each hinge assured deployment and increased loads on the stops. The normal spring end and coil forces push each hinge pin against its clevis holes to prevent hinge rattle.

2.3. Magnetometer Boom Design

Carbon fiber/epoxy was chosen as the primary structure of the booms due mainly to its very high stiffness and strength to weight ratio, as well as the low thermal expansion coefficient. The magnetometer signal and power wires were routed inside the boom tubes and hinges using coaxial cables. Each boom also included an outboard connector, so that the magnetometers could easily be removed for testing. Harnessing was reduced to about 40% of the customary weight by using Goretex dielectric and helically served foil coaxial shields with six drain wires. Served shield coax is not only lighter and more flexible than the braided shield option, but also provides better electrical coverage. In spite of these weight saving measures in the electrical harness, it was still 25% heavier than the composite boom tubes. The total mass of the two booms, including the restraint mechanisms, saddles, tubes, balancing mass, and interfaces to the magnetometers is 3.5 kg. The total mass of the three magnetometers is 1.8 kg.

The carbon fiber/epoxy tubes were constructed from 5 layers of woven prepreg material. The 0.14 mm thick layers were laid out with 0°/90° orientation for the outer, inner, and middle layers, and ± 45° orientation for the two mid-layers. The finished tubes have an outside diameter of 32 mm and a wall thickness of 0.7 mm. The tubes were fabricated by SSL personnel in the Composite Materials Laboratory of the UC Berkeley Mechanical Engineering Department, directed by Prof. Hari Dharan. The epoxy pre-impregnated material was laid by hand on a polished aluminum mandrel, then wrapped with a film of PTFE Teflon. Instead of using an autoclave or vacuum bagging, a length of thick-walled neoprene heat shrink tube was placed around the uncured tube. The shrink temperature of the tubing is the same as the cure temperature of the epoxy in the prepreg (175 °C). The entire assembly was baked for two hours and cooled. The shrink tube was carefully sliced off, and the aluminum mandrel thermally contracted to release the finished tube.

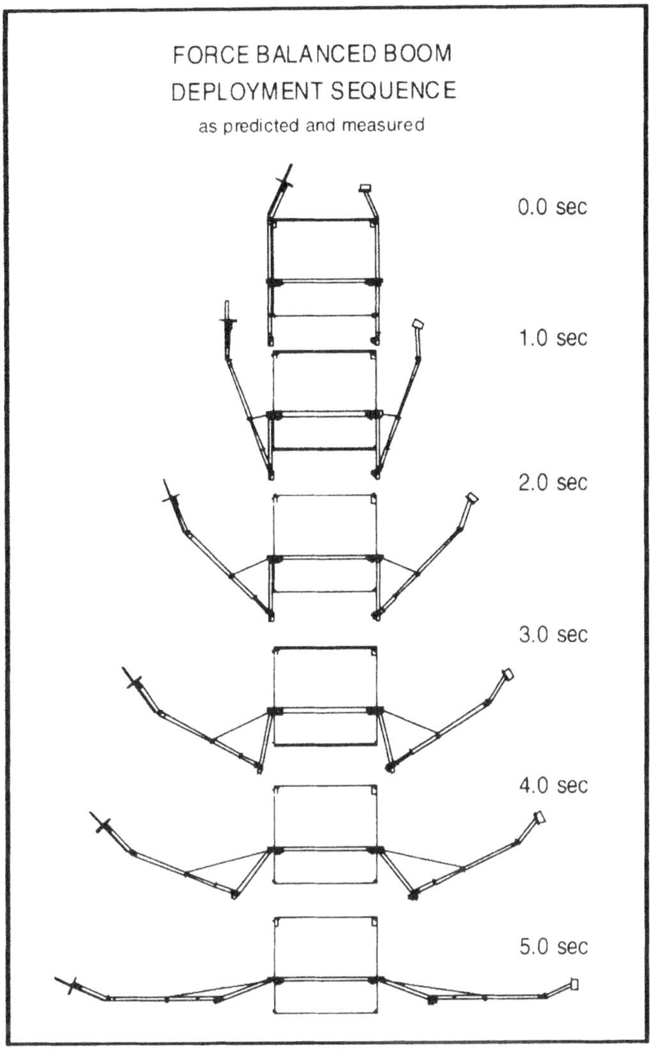

Figure 3. Magnetometer Boom Deployment.

Avoiding the use of vacuum bags or autoclaves significantly reduced labor and expense and produced uniform, high quality tubes. The metal to composite tube joints used thin aluminum sleeves bonded with structural epoxy (both internal and external) to minimize both distortions and stress concentrations. Alternating bands of aluminum tape and conductive black tape were chosen by the SMEX office for the thermal control, so as to minimize the temperature excursions in shadow.

2.4. Magnetometer Boom Testing

Testing of the boom systems included vibration and deployment testing. Deployment tests were conducted in two configurations: ceiling drops and spin tests.

For ceiling drop tests, a boom with magnetometer mass dummy was installed on a fixture simulating the side of the spacecraft mounted to the ceiling of the test facility. The fixture was tilted about the spin axis to approximate tangential coriolis accelerations of the actual deployment. The boom was released, and the boom segment relative angles were measured using potentiometers at the hinges. Overload tests were conducted using oversized instrument mass dummies. Spin tests were conducted on a spin table with a fixture simulating two sides of the spacecraft. Also mounted to the table was an overhead track on which rode trolleys carrying constant-force springs. The springs were fastened to key points on the magnetometer booms to counter the effects of gravity. The table was spun at various speeds, and the booms were deployed to verify computer simulations. Numerical prediction of the deployment environment was a vital part of the design process, to establish the needed strength and stiffness margins.

Deployment testing and flight deployments showed that the final position of the magnetometers were repeatable to $\sim$1 mm, while the science magnetometer orientation was repeatable to $\sim$3 arc min, and deployed boom resonance was above 2 Hz. The unequal weights of the three magnetometers led to concerns of spacecraft balance. Balance weights were added such that that the spacecraft was dynamically balanced with the booms removed, or deployed. As a result, FAST was launched with an intentional one degree tilt of the spin axis, which went to zero when the booms were deployed shortly after separation.

3. The Radial Electric Field Mechanisms

3.1. The Sensors

The primary sensing elements for the FAST vector electric fields experiment are ten 8 cm spherical sensors. Each axial boom has one sensor, while each of the four 30 m wire booms has two sensors separated by five meters. Each of the sensors contains a pre-amplifier which uses a variety of active feedback schemes to achieve an ultra-high impedance measurement of the local plasma potential, while also minimizing the local disturbances of the plasma. The preamps are housed in 1 mm thick aluminum shells, which are coated with electrically conductive DAG-213 (an epoxy based paint doped with fine carbon particles).

The ∅2.5 mm electrical cable used for wire booms is of an advanced construction, designed at UCB/SSL and fabricated by W.L.Gore, Inc. This cable includes two ∅0.6 mm fifty Ohm coax cables and eight AWG #36 conductors. The central wire bundle is jacketed with a Kevlar braid, an aluminized Kapton conductive foil, and a silver plated copper outer braid. With Kevlar as the internal strength member of the cable, we were able to cut and electrically isolate the outer braid, as desired. In fact, each boom cable is divided into eight surfaces that are electrically biased, so as to minimize the disturbances of the conductors and spacecraft in the local

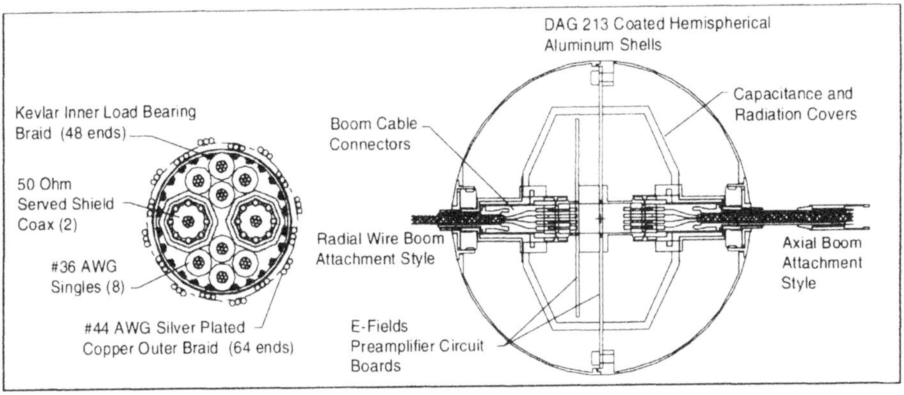

Figure 4. Wire Boom Cable (20×) and Electric Fields Spherical Sensor.

plasma. Special attention to design and fabrication details ultimately provided a very flexible cable, which was shown to deviate less than 10 mm from the desired limp wire radial line. One of the formidable engineering tasks was to also produce a very low thermal expansion cable, so as to avoid 'wind-up' effects of the rapidly cooled and reheated cable each time the satellite traverses the earth's shadow. The Kevlar braiding could not alone provide the desired low expansion cable. We utilize the fact that the cable core is mostly Teflon insulation (by volume) covered with a helical wrap. This results in an apparent negative expansion as the helical path length increases (when the cable warms in sunlight) due to an expanding Teflon core diameter. The sphere connectors provide a 'plug & play' design, wherein any of the cables, sensors, or preamps could be rapidly swapped out (as needed) without mechanism tear down or cable de-soldering. The ∅10 mm × 15 mm miniature boom cable connectors for each cable end were also a custom UCB design using developmental pins and sockets provided by Hypertronics, Inc.

A major boom mechanism design challenge was the proper handling of the spheres. Since the painted surfaces of the spheres are the primary science sensing surfaces, blemishes or abrasions simply could not be tolerated. No contact of the sphere surface could occur during ground handling, launch, or deployment. Small ∅4 mm stubs protrude from either side of each sphere, and these were used to support the sensors.

3.2. Radial Wire Boom Deployment and Configuration

Each radial wire boom has two spherical sensors and three cable assemblies; a 23 m inner cable, a 5 m sphere to sphere cable, and a 2.5 m 'tail wire' (extending beyond the outer sphere to provide electrical, thermal and mechanical symmetry about the sensor).

A mechanism design issue was the on-orbit activities needed to deploy this rather complex wire geometry, without impacting the delicate body mounted solar array, touching the surfaces of the spheres, or tangling the cables on the magne-

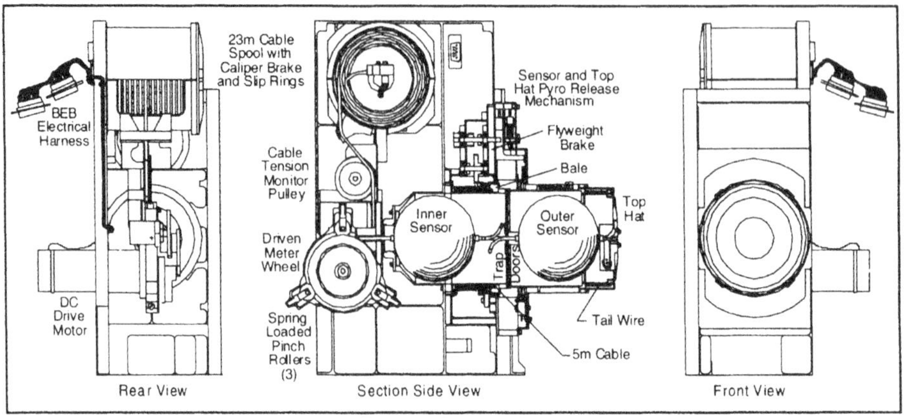

Figure 5. Wire Boom Deployment Mechanism.

tometer booms. The deployment of the wire boom was broken up into three events corresponding to each cable assembly. The three cable assemblies were deployed by a different method.

3.2.1. *5 m Cable Deployment*

The outer sphere and a protective 'Top Hat' were released by the firing of a pyro, and pulled outward by an initial centripetal acceleration of 0.3 G. The centripetal acceleration varies as the outer sphere travels away from the spacecraft, and increased to over 1.5 G. This first deployment of a pair of booms also slowed the spin of the spacecraft from 25 to 17 rpm, within approximately 60 s. A centrifugal brake on the deploying cable limited the cable feed rate to below 0.08 m s^{-1}, in order to keep the coriolis induced wire boom offsets to less than 10° from true radial, well away from the magnetometer booms, which are at 45° from the wire booms.

3.2.2. *Tail Wire Deployment*

After an opposed pair of 5 m sphere to sphere cables were deployed, the spacecraft was spun back up to 25 rpm to release the Top Hats. The Top Hats were held to the outer sphere by a force latch that released at ~2 G's. Release of the Top Hats rapidly unfurled the $2\frac{1}{2}$ m tail wire, well away from the spacecraft. The Top Hats were discarded.

3.2.3. *23 m Cable Deployment*

With the outer sensors and cables deployed, the wire booms were ready for the third event, metered deployment of the 23 m main cable. On orbit, this stage was done in two meter increments, with constant ACS spin-up torqueing, in order to maintain a minimum 12 rpm spin for continuous science data.

3.3. Deployment Mechanisms

3.3.1. *Top Hat Release*

For launch, the inner sphere was housed inside a spool within the body of the boom mechanism. The outer sphere was housed in the Top Hat, which was held against the boom housing by four latches. The outer $2\frac{1}{2}$ m tail wire was wrapped into an annulus in the Top Hat. A ⌀0.61 mm hat retention cable circled the Top Hat, and held the latches against the Top Hat with a noose-like action. The cable circle changes diameter, and correspondingly drives in or releases the latches, with the position of a tension arm. The tension arm position was also used to simultaneously lock and unlock the centrifugal brake mechanism. The tension in the hat retention cable was maintained with a pair of redundant springs, which were tied to the tension arm by a pair of ⌀0.46 mm cables. Release of the Top Hat and flyweight brake occurred with the firing of a pair of redundant pyro operated cable cutters, each of which severed the pair of ⌀0.46 mm cables, freeing the tension arm, allowing the hat retention cable circle to expand, thereby releasing the latches. Confirmation of Top Hat release is provided by a microswitch acting directly on the edge of the Hat. The pyrotechnics are ICI Americas type 192Y subminiature wire cutters. They are ⌀4.8 mm × 27 mm and contain a 25 mg explosive charge.

3.3.2. *5 m Cable Payout*

The sphere stubs between sensors were supported by a pair of spring loaded trap doors. The 5 m sphere to sphere cable was routed to the exterior of the spool, and wrapped into an annulus around it. The cable was held in place simply by fixing the cable endpoints. A rotating 'bale' that will later allow the cable to peel out from the end of the spool, one coil at a time, lay at the external edge of the spool. This bale operation is analogous to, but opposite of, the bale on a spin casting fishing reel. In the wire boom, a rotation of the bale is required to remove cable from the spool, with the fishing reel, the bale rotation acts to wind fishing line onto the reel. The bale is linked to a centrifugal (flyweight) brake through a 20:1 step up gear train, to control the speed of cable pay out. The flyweight brake uses spring loaded PEEK friction pads, acting against an aluminum housing. The entire gear train is supported by Barden dry film lubricated ball bearings. No preload of the bearings can be tolerated, as the bale must function with only a 5 g tangential input. All gear and bearing spacing requirements are therefore controlled by shims. This was the most critical aspect of the mechanism, as even a light preload from wavy washers caused undetected imperfections or perhaps dry film particulate in the bearings to produce unacceptable spikes in the rotating torque. The brake mechanism maintains a payout speed of 0.06–0.08 m s^{-1}, despite variations in the wire tension of over a factor of 5.

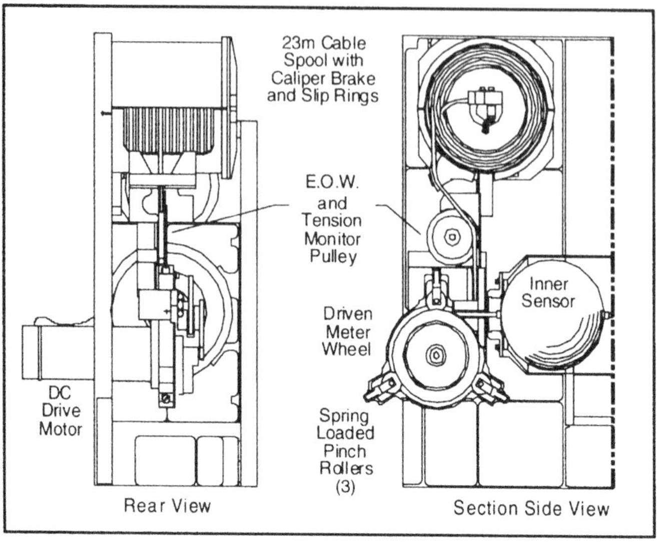

Figure 6. Spool Mechanism Detail.

3.3.3. *2.5 m Cable Payout*

The release of the Top Hat by the latches freed the Hat, which contains a spring loaded Acetal (Delrin) sleeve. This sleeve pushed the Top Hat away from the spacecraft, and simultaneously opened the end of the annulus containing the 2.5 m cable. Constraint of this cable is required for launch, and the sleeve motion allowed it to peel out of the annulus freely, upon separation of the sphere from the Top Hat. The force latch which controlled the separation of the sphere from the Top Hat consists of a torsion spring acting on an arm, which engaged the sphere stub through a small hole in the stub. The preload of the torsion spring was set such that a 2 G load on the Top Hat will allow the arm to rotate, which released the sphere.

3.3.4. *23 m Cable Payout*

The 23 m cable deployment mechanism (see Figure 6) contains a rotating cable storage spool, a metering wheel cable deployment assembly, a vacuum-service brush DC gear motor, over-tension and end-of-wire (E.O.W.) indicators, analog and digital cable length indicators and redundant gold alloy slip ring electrical contacts on the inboard end. The mechanism deploys cable at a nominal rate of 8 mm s^{-1} with tension ranging from 0 to 100 N. The drive motor is magnetically and EMI shielded. The meter wheel cable grip is enhanced by three equally spaced spring loaded pinch rollers. The meter wheel and pinch rollers have fluorocarbon (Viton) surfaces to avoid damaging the cable braid. The internal cable over-tension indicator prevents breaking and 'discarding' the cable, should a motor malfunction occur.

3.4. WIRE BOOM TESTING

Both horizontal and vertical deployments were used to test the wire boom deployment, most testing being focused on the dynamics of the 5 m cable segment. Vertical deployments downward were used with varying masses at the sphere/Top Hat location to verify margins and to allow 3-dimensional dynamics. Deployment was demonstrated with masses of 39, 119, 350, 606, 2270 and 3180 g. This demonstrates a 0.11 to 9 G range, simulating a range of spacecraft rpm of 15–135. The predicted 8° coriolis angle was simulated by tilting the boom housing during these deployments.

Horizontal tests both in air and in thermal vacuum were conducted by constraining the sphere/Top Hat to a linear track, and allowing the boom to pivot freely about it's center of gravity. This only provides one plane of motion for the deployment dynamics, but was sufficient for acceptance testing of the flight units, allowing operation through the range of −40 °C to +60 °C to be demonstrated. All horizontal testing employed a continuously variable clutch to provide a tensile force on the Top Hat in place of the centripetal force. This force could be changed during the deployment to match the force profile that would be encountered on orbit, but in practice, the lowest level initial force was retained throughout the deployment as a worst case. The subsequent spin up and Top Hat separation force were verified by increasing the tensile force on the Top Hat with the variable clutch until separation occurred.

Each boom was deployed horizontally after 3 axis vibration testing, and after thermal/vacuum cycling.

4. The Axial Electric Field Mechanisms

4.1. DESIGN AND CONFIGURATION

The axial booms consist of the boom element (Stacer), guiding and stiffening nozzles, a deployment assistance device (DAD), spherical sensor with electric field biasing elements, and a carbon-fiber housing.

4.2. STACER

The FAST spin axis sensors use a boom commercially known as a 'Stacer', a proprietary device developed by the Hunter Spring Co, and now owned by Ametek. The boom is a rolled strip material of spiral-wound, fixed-helix angle that forms a gently tapered hollow tube when deployed. In its stowed state, it fits within a short cylinder. This boom geometry resembles the 'Chinese yo-yo' sometimes used as birthday party favors.

The Stacer element is cold formed using 0.12 mm thick × 126 mm wide Elgiloy strip material. The strip is formed with a constant (fixed) free coil diameter and

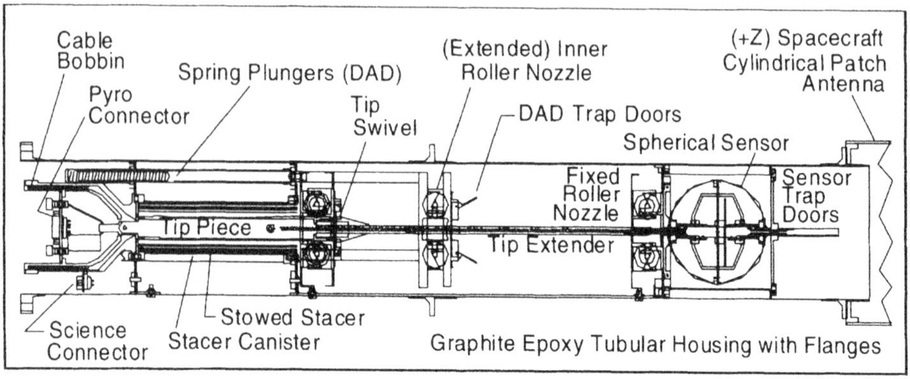

Figure 7. Stowed Axial Boom (section view).

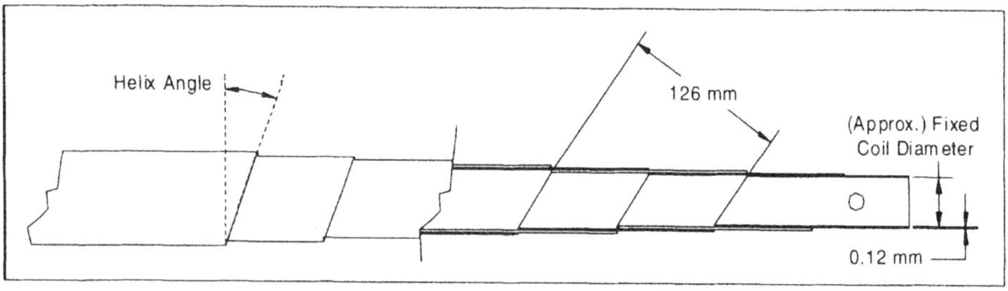

Figure 8. Stacer Detail.

helix angle. The preferred forming geometry involves significant coil overlap, such that two layers of strip material would be found at any point along the length of the extended Stacer element. The deployed end inner-most coil cinches on a cylindrical tip piece and the subsequent coils of an extended boom element 'stack' on the previous coils producing the characteristic taper. Typical deployed coil diameters are slightly larger than the free coil diameter, which means that normal forces exist between coils due to this 'stacking' process. These intercoil forces in turn produce intercoil friction forces that prevent gross coil slippage. The resulting boom has bending stiffness and strength which compare favorably with the equivalent solid thin walled tube.

A primary advantage of the Stacer element, as compared to other stiff boom geometries, is its thermal symmetry. The usual alternative uses two longitudinal cambered (>180°) strips that overlap lengthwise to form a tube. The low thermal conduction between strips can produce significant thermal bowing (Staugaitis and Predmore, 1973), which has been known to cause thermal pumping on slow spinning spacecraft. The helical overlap of the single Stacer strip, by contrast, is axi-symmetric for circumferential heat flow. Other advantages of the Stacer are that it does not need a motor driven deploy mechanism, and the fact that a cable may easily be fed up the center of the boom.

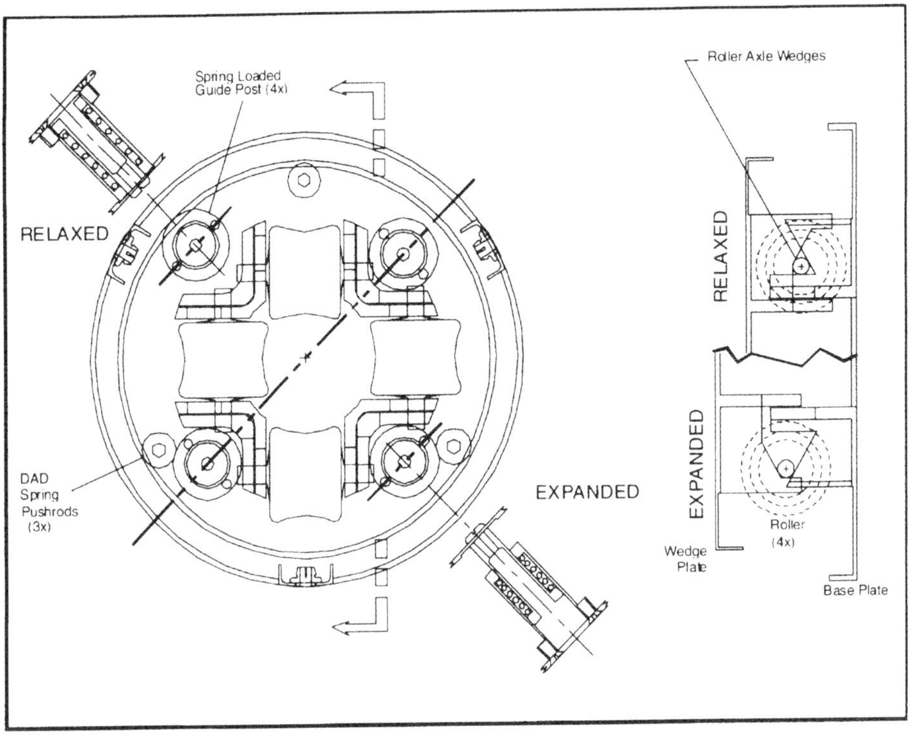

Figure 9. Axial Boom Roller Nozzle Detail.

4.2.1. *Boom Support Nozzles*

Two roller nozzles were used to give the axial boom the required cantilever stiffness. They were positioned beyond the Stacer canister: one attached to the deployment assistance device mounted on the canister itself, and the other mounted on the end of the main carbon fiber housing. This design maximized the separation between the nozzles to provide a rigid pinned-pinned base support.

The expanding roller nozzle design was developed to accommodate the changing boom diameter, while also minimizing friction. Axles of the four rollers are held between the base and wedge plates by spring loaded guide posts. The resulting geometry allows concentric outward movement of all rollers, but produces a friction lock when only one roller is loaded by cantilever boom bending, thus providing a centering force. The boom element thus has full cantilever stiffness.

4.2.2. *Deployment Assistance Device (DAD)*

The inner roller nozzle was held against the canister for launch by trap doors that caught the edge of the Stacer tip piece, and was then pushed beyond the coil transition zone by spring loaded plungers when the boom was released. These plungers also provided an initial kick force to help the 'cinching' of the first coil on the tip piece, and a robust deployment of the remaining coils. After the pyro initiated release, the DAD plunger pushed the tip piece to beyond the coil forming zone of the Stacer, and then the trap doors were pushed open by the deploying boom. Once the boom was mostly deployed, the rollers were contacted, and limited the booms range of motion to its axial component.

4.2.3. *Electric Field Biasing Elements*

As mentioned before, the best e-field measurements can be made if the photo-emissions from the space craft and the boom mechanisms can be nullified. The axial booms used stiff stubs that were insulated from the main spherical sensor and the Stacer by PEEK plastic. Thin-walled aluminum tubes: ⌀ 6 mm × 250 mm long were used for these stubs, coated with the same DAG 213 as the sensors. These provided the electrically bias able inner 'stub' segments. A shorter piece of the same material was mounted to the exterior of the sensor, for symmetry.

A UCB design electrical connector between sensor and stub allowed swapping of the sensors without disturbing the boom. This assembly was then attached to the Stacer tip piece, utilizing a rotating sleeve in a collar that allows the Stacer to deploy without causing the sensor cable to twist. The edge of the collar projected to form a lip for the DAD trap doors to catch.

The axial boom electrical cable was stored for launch on a bobbin below the boom canister. The bobbin annulus was sized to prevent crossing and subsequent tangling of the cable coils. The cable was pulled from the end of the bobbin, drawn up the center of the boom element as it deployed, and also aided in stopping the element at its desired final length. A microswitch was positioned to record the unwinding of each coil as it was pulled off the bobbin, giving length and sensor velocity measurements.

4.2.4. *Carbon Fiber Housing*

The entire axial boom mechanism was secured to the FAST spacecraft inside carbon fiber housings, one each for $+Z$ and $-Z$ axes. These housings were built on the UCB campus in the Composites Lab alongside the magnetometer booms.

Each boom assembly is mounted in a ⌀100 mm thin wall graphite/epoxy tube. The tube consisted of 5 layers of a Fiberite 0.14 mm thick prepreg woven carbon material. The inner, middle and outer layers were laid-up at 0°/90° while the layers between were set to ± 45°. They were laid up on a polished aluminum mandrel, covered with perforated Teflon release material, bleeder batting and vacuum bagged before curing. After curing, the thermal contraction of the aluminum mandrel as it cooled allowed the tube to slip free. The lower tube was 670 mm long, the upper tube was extended to 800 mm in order to support the S-band antenna well away from spacecraft. After light bead-blasting to remove the insulating epoxy surface, flanges were epoxied to the outside of the tube to provide mounting points for thermal blankets, the antenna and the flange to bolt the boom to the S/C deck. This bead-blasting also allowed the tube to meet electro-static cleanliness requirements without further processing, by exposing the carbon fibers.

4.3. Axial Boom Testing

The testing of the axial booms utilized 2 methods of deployment. Initially the boom was deployed vertically upward to yield an unbiased-by-gravity concentricity/target accuracy measurement. The unit was mounted coaxially on a spin-table base, and after deployment, the unit was spun and the total indicated run out (TIR) was measured. The axial boom was then deployed horizontally with the tip piece attached to a low friction trolley on a track. This allowed the Stacer to develop full speed, which gave full length, and therefore allowed a very accurate measurement of deployed length. The deployment microswitch state-changes were measured for all deployments and compared. 3 axis non-combined vibration tests were run, after which the booms received 4 hot/cold cycles of thermal vacuum. The horizontal deployment rig was also used in thermal vacuum testing: after hot/cold soaking, each boom was deployed down the trolley track in a 'snout' attached to the chamber. One boom was deployed hot (+50 °C), the other cold (−35 °C). After inspection, a final verification deployment was done. Final inspection and stowing for launch completed the axial booms.

Deployed boom straightness was found to be the order of ±1 cm at its 3.8 m length, and the cantilever resonance was above 1 Hz. In this free-pop configuration, the extension velocity reached ~6 m per second. As such, it was prudent to include some form of extension restraint in order to reduce the shock forces at the end of the extension stroke. This was achieved by sheathing the last bobbin coil with PTFE shrink tube. The tight fit within the bobbin annulus provided the needed retarding force as the boom reached full length. Without the retarding force, there exists a chance that the boom momentum could tear out the retaining rivet, causing a 'floppy' boom.

4.4. Deployment

Prior to deployment, the Stacer element was elastically compressed into the canister whose length is only the strip width. The DAD was also compressed, and the two sets of trap doors placed in their launch positions. The outboard passive 'trap doors' supported the sphere for launch. The sensor cable was then wound around the bobbin, and carefully slid into it's housing. Last, the pyro assembly was installed and tensioned. The boom was restrained by a 'fishhook' geometry on the rear of the tip piece. This hook was held with an aircraft cable loop that was cut by pyro activated cable cutters to release the boom.

Upon signal from the ground, the pyro ignited, and cut the wire. The Stacer then rapidly self extended by its spring energy as a one shot device, aided by the DAD. The spring force is due to the release of strain energy as the larger diameter stowed coils 'collapse' onto the previously formed boom coils within a single coil transition zone. At full extension, the boom was positioned by the 2 roller nozzles resting tightly against the Stacer element.

5. On Orbit Performance

The FAST spacecraft was launched on a Pegasus XL on August 21, 1996. Within one week, the spacecraft was turned on, and the magnetometer booms deployed at a spin rate of 20 rpm. After a spin up to 25 rpm, the deployment sequence was begun on the first pair of wire booms.

One of this first pair of wire booms, boom #2, stopped after 25 cm of deployment, the other deployed out to 5 m. In an attempt to free the stuck boom and to release the Top Hat from the sphere, the spacecraft was spun up to 40 rpm. Subsequently, the spacecraft was slowed down, then 'kicked' with the highest impulse available with the torque coil. None of these maneuvers were able to move the boom any farther out. The second pair of wire booms were then deployed, both extending to 5 m. Top Hat separation occurred during the 5 m deployment.

The three booms that deployed to 5 m were then extended out to their full 23 m length. Because the booms deployed from slightly below the CG of the spacecraft, the asymmetric load due to the partially deployed boom caused the spacecraft to tilt. Deployment of the $+Z$ axis axial boom shifted the spacecraft CG upward, which decreased the tilt. Because deployment of the $-Z$ axis boom would shift the CG farther below the plane of the wire booms, it was decided not to deploy this boom.

The on orbit deployment failure in wire boom #2, most likely due to the 5 m cable crossing over itself during launch, was unfortunate. The location of this boom during the horizontal Pegasus launch was down, superimposing a DC acceleration during the launch and pull-up which would tend to bias any cable motion to the edge of the Spool, where the possible cable crossover would have occurred.

This has moved the composite center of mass $\sim$5 cm from the spacecraft axis, causing the two adjacent wires to shift 6° in the spin plane, toward the missing boom. This somewhat triangular wire boom orientation still provides excellent planar components of the vector fields measurement. The vehicle spin axis has only a 0.4° tilt from the geometric axis, well within the established science tolerances. This is also flight verification of the widely held opinion that vehicle dynamic balance takes precedence over the static balance.

The unprecedented science data obtained is a tribute to the robust nature of this system architecture. On orbit, the FAST vehicle has been found to have excellent attitude stability and spin axis maneuverability. Eclipse spin up, caused by rapidly cooling wires, has been found to be only one part in several thousand. The 'negative orbit normal cartwheel' orientation of spin to orbit makes the vehicle relatively immune to gravity gradient disturbances.

6. Summary

The FAST boom mechanisms designs have evolved over a period of twenty years, from a series of successful satellite instruments flown on S3-3, ISEE, Viking, Freja, CRRES, Polar, and Cluster 1 and 2. This FAST mission was a unique opportunity to advance the state of the art of the three boom systems it employed. It was predicated as a higher risk mission with a short development and fabrication cycle. The spacecraft subsequently achieved its science objectives, and by doing so, provided engineering development history for future missions to call upon.

Acknowledgments

Dr. Charles Carlson, University of California Berkeley is Principle Investigator for the FAST Mission.

References

Lai, S. T. and Bhavnani, K. H.: *Dynamics of Satellite Wire Boom Systems*, AFCRL-TR-75-0220.

Meirovitch, L. and Calico, R. E.: 1972, *The Stability of Motion of Satellites with Flexible Appendages*, NASA CR-1978.

Staugaitis, C. L. and Predmore, R. E.: 1973, *Thermal Static Bending of Deployable Interlocked Booms*, NASA TN D-7243.

Thomson, W. T.: 1961, *Introduction to Space Dynamics*, J. Wiley, New York.

THE FAST SPACECRAFT INSTRUMENT DATA PROCESSING UNIT

P. R. HARVEY, D. W. CURTIS, H. D. HEETDERKS, D. PANKOW,
J. M. RAUCH-LEIBA, S. K. WITTENBROCK and J. P. McFADDEN
Space Sciences Laboratory, University of California, Berkeley, U.S.A.

Abstract. The Fast Auroral Snapshot Explorer (FAST) is the second of the Small Explorer Missions which are designed to provide low cost space flight opportunities to the scientific community. FAST performs high time resolution measurements of the auroral zone in order to resolve the microphysics of the auroral acceleration region. Its primary science objectives necessitate high data volume, real-time command capability, and control of science data collection on suborbital time scales. The large number of instruments requires a sophisticated Instrument Data Processing Unit (IDPU) to organize the data into the 1 Gbit solid state memory. The large data volume produced by the instruments requires a flexible memory capable of both high data rate snapshots ($\sim$12 Mbit s^{-1}) and coarser survey data collection ($\sim$0.5 Mbit s^{-1}) to place the high rate data in context. In order to optimize the science, onboard triggering algorithms select the snapshots based upon data quality. This paper presents a detailed discussion of the hardware and software design of the FAST IDPU, describing the innovative design that has been essential to the FAST mission's success.

1. Introduction

The Fast Auroral Snapshot (FAST) is the second of the Small Explorer (SMEX) missions. The FAST mission is designed to measure the microscale physics of the auroral acceleration region with the highest time resolution possible. A sophisticated Instrument Data Processing Unit (IDPU) manages the entire science package, collecting and organizing the data before transmission to the ground. The IDPU controls a large number of instruments: 16 electrostatic analyzers to measure charged particle fluxes, a mass spectrometer, 10 electric field and 6 magnetic field sensors. In addition, support electronics include electric and magnetic field sensor signal conditioning and analysis, particle data averaging, and housekeeping monitoring. Since it was known that the envisioned data rates combined with instrument control could easily overwhelm a processor, a design philosophy was chosen to develop autonomous instruments that require little commanding and a data recording system that minimizes processor interaction. Numerous tasks normally performed by the processor have been delegated to custom design circuits using Field Programmable Gate Arrays (FPGAs). With a reduced work load for the processor, the software focuses on data optimization schemes that maximize the science return.

The high data accumulation rate ($\sim$12 Mbit s^{-1}) required during scientifically interesting events generates a data volume far too large for collection (a 40-min

Space Science Reviews **98:** 113–149, 2001.

auroral crossing generates ~3600 MBytes), too large for transmission (a typical 20-min contact at 2.25 Mbit s^{-1} allows only ~340 MBytes), and too large for conventional data recorders available at the time FAST was built. The science requirements drove the development of a solid state recorder whose 125 MByte (1 Gbit) size was limited by mass constraints. The recorder was included in the IDPU to eliminate a high bandwidth interface between the spacecraft and the IDPU, and to allow more control of the data recording by the IDPU. The IDPU formats data directly into packets and frames so that it is ready for transmission without further processing. Design specifications required the IDPU to simultaneously record and play back data from several different asynchronous channels, prioritizing the high resolution data for selective transmission to the ground. In order to place the high resolution data in a large scale auroral context, coarse survey data collection was required throughout the auroral zone. To accommodate different data types, the FAST team developed four levels of data collection: slow survey (<0.05 Mbit s^{-1}), fast survey (~0.5 Mbit s^{-1}), burst (5-12 Mbit s^{-1}), and high speed burst (80 Mbit s^{-1}). The IDPU controls the survey data collection rates based upon trigger data, and prioritizes high resolution bursts (5–30 s duration) for selective transmission to the ground. This data optimization scheme has led to high quality, scientifically interesting data being collected on nearly every orbit.

This paper presents a detailed discussion of the hardware and software design of the FAST IDPU. The IDPU consists of a processor board, formatter board, solid state memory, and power system, plus electric and magnetic field signal conditioning and analysis circuits described in Ergun *et al.* (2001). The processor is the primary interface to the spacecraft Mission Unique Electronics (MUE, a dual processor electronics package that monitors spacecraft health, controls telemetry, receives and relays commands, performs attitude control, and manages power, see Pfaff *et al.*, 2001). The IDPU controls and routes commands to the various instruments and IDPU electronics boards, controls the science power system, compiles housekeeping data, directs traffic to the memory, generates packet headers, and manages the onboard data selection using triggering algorithms. The processor also performs one-time tasks such as wire boom deployments (Pankow *et al.*, 2001). The formatter board acts as a router directing data from the particle instruments (Carlson *et al.*, 2001; Klumpar *et al.*, 2001) to the memory. It also forms survey averages of the particle data, compresses the data, and generates particle trigger inputs that are passed to the processor. The solid state memory is a 125 MByte error-corrected high-speed RAM that can be configured to store the multiple data types (survey data, burst data, etc.) at data rates up to 12 Mbit s^{-1}. The power system is designed to provide stable, regulated voltages and contains a convenient power switching relay system that efficiently reduces instrument power when data is not being collected. The innovative design of the FAST IDPU has been essential to the mission success and can be considered a model for integrating multiple spacecraft instruments into a single experiment.

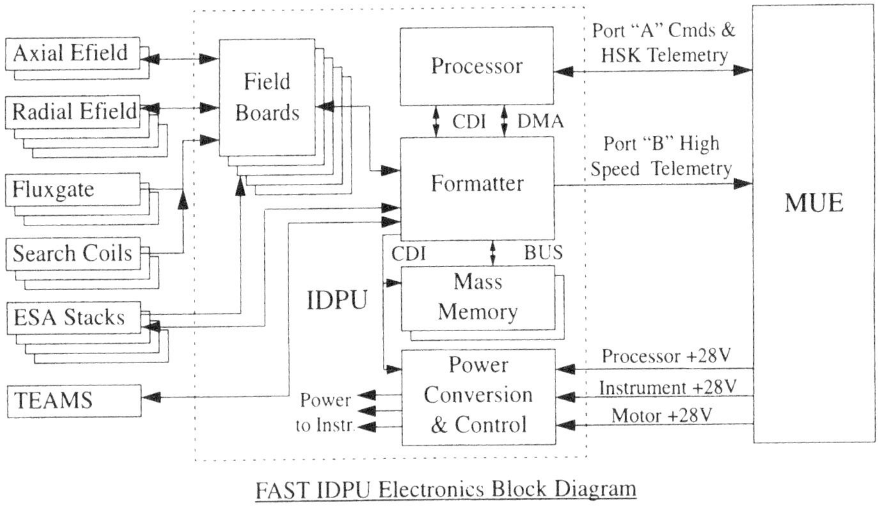

Figure 1. Block diagram of the FAST Instrument Data Processing Unit (IDPU) and interface to the spacecraft Mission Unique Electronics (MUE).

This paper is organized into two main sections covering hardware and software. The hardware section includes descriptions of the processor, formatter, memory and power subsystems. These subsystems form a single, compact electronics box that is housed near the center of the spacecraft to maximize radiation shielding and reduce harnessing. The software section examines the operating system, engineering subroutines, command management, telemetry management, and instrument control. The flight software was written entirely in assembly language to reduce its size and to optimize its performance, and requires a mere 31 KBytes of code space and 60 KBytes of RAM.

2. Hardware Description

The IDPU includes a Processor, Formatter, Mass Memory, and Power section, plus Electric Field and Magnetometer signal conditioning and analysis circuits (see Figure 1). The Processor card provides the intelligence of the system, controlling the flow of raw data through the Formatter and Memory system. The Formatter card performs data collection, averaging, formatting, storage into the memory, and transmission from the memory to the spacecraft with minimal Processor intervention. The Memory system provides 125 MBytes of error-corrected memory used for science and engineering data storage. The Power system provides DC/DC conversion, power distribution, current limiting and monitoring. The Electric and Magnetic Field conditioning circuits are described in a separate paper (Ergun *et al.*, 2001).

IDPU circuitry is packaged in a single box to reduce mass, simplify harnessing and interfaces, and reduce duplication in power converters and other common ser-

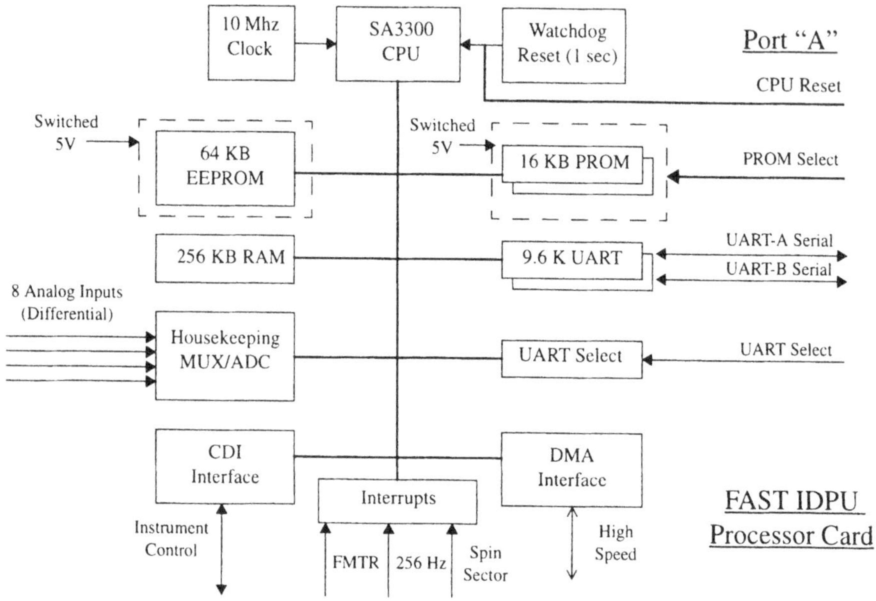

Figure 2. Block diagram of the FAST processor board.

vices. The IDPU makes heavy use of Field-Programmable Gate Arrays (FPGAs) to improve density over a small scale integration (SSI) logic based system, and speed over a processor-based system.

Electronic parts for the IDPU were selected according to criteria for functionality, reliability, and radiation hardness. Moderate reliability levels (e.g., Mil-STD-883C for microcircuits) were coupled with extensive instrument-level testing to verify reliability. Parts were specified and/or lot tested by the FAST team to achieve a minimum tolerance of 20 Krads, and shielded as necessary to 100 Krads. Parts with Single Event Upset (SEU) sensitivity were protected against latchup by current limiting circuits, and against bit-flips by Error Detection and Correction (EDAC) logic.

2.1. Processor

The Processor system is based upon a Sandia SA3300 32-bit processor, operating at 10 MHz. As shown in Figure 2, the processor card features redundant sections of 16 KByte boot PROM, 64 KBytes of EEPROM, and 256 KBytes of RAM. The card includes a housekeeping A/D system, watchdog timer, serial I/O to the spacecraft MUE, and special interfaces to the Formatter card.

The processor and local memory are rad-hard and SEU immune. The PROM is bipolar, which makes it rad-hard and fast, but power hungry. To save power, the PROM is turned off after being copied to RAM during power-on initialization. Two copies of the PROM code exist, selectable from the spacecraft interface to provide redundancy.

The EEPROM is also power-switched and isolated to save power and to allow the system to operate in case of an EEPROM failure. (The EEPROM is somewhat less rad-hard than the rest of the design, although it meets the mission requirements.) EEPROM is used to hold additional code which can be modified by uplink memory loads after launch as needed. After building the code from the PROM and EEPROM images at power-on, the code is executed out of RAM.

The Processor also includes a 10-bit low-power 'Flash' Analog to Digital Converter (ADC) for digitizing the instrument analog housekeeping. The converter is connected to a tree of analog multiplexers, some on the processor and some remote in the instruments. The multiplexers are controlled by the processor to allow the processor to sample a large variety of housekeeping signals with differing sample rate requirements.

A pair of RS232-style serial interfaces, or Universal Asynchronous Receiver Transmitters (UARTs), on the Processor provide a redundant bi-directional interface between the IDPU and the spacecraft MUE. This interface is used for low rate data (9600 baud) such as commands, status, housekeeping, timing, and attitude control information. A spacecraft-controlled line selects which UART to use.

The Processor communicates with the Formatter and all instrument electronics using a custom bi-directional serial Command and Data Interface (CDI). The CDI transfers data at 1 Mbit s^{-1} in 24-bit words, which include an 8-bit destination address and a 16-bit data value. The Formatter and instruments monitor this interface, decode the register address, and latch information sent to them. When Formatter status registers are addressed, the data direction of the serial interface is reversed, and the 16-bit data value is returned to the Processor.

A second serial interface between the Formatter and the Processor transfers large blocks of data between the Processor and the Memory system (via the Formatter) at 4 Mbit s^{-1}. This interface is called the Direct Memory Access (DMA) interface, and is used primarily to transfer engineering packets and science packet headers into the memory for later transmission.

2.2. Formatter

The Formatter board (10 FPGAs and some memory chips) was designed to perform processor-like tasks at high speed. The Formatter card interfaces the IDPU with the FAST instruments over twenty separate serial interfaces (see Figure 3). Some of these instrument interfaces provide spin-synchronous data, others provide time-synchronous data, and some provide data only when requested by the Processor. In addition to these data sources, the Formatter produces various averages of these data for lower telemetry rate 'Survey' data sets, bringing the total number of data sources to twenty-nine. Data from each source is formatted into its own 1024-byte frame and transferred into the Memory. The Processor adds CCSDS (Consultative Committee for Space Data Systems) telemetry frame and packet headers (one packet per frame) using the DMA interface. Each data source has its own desti-

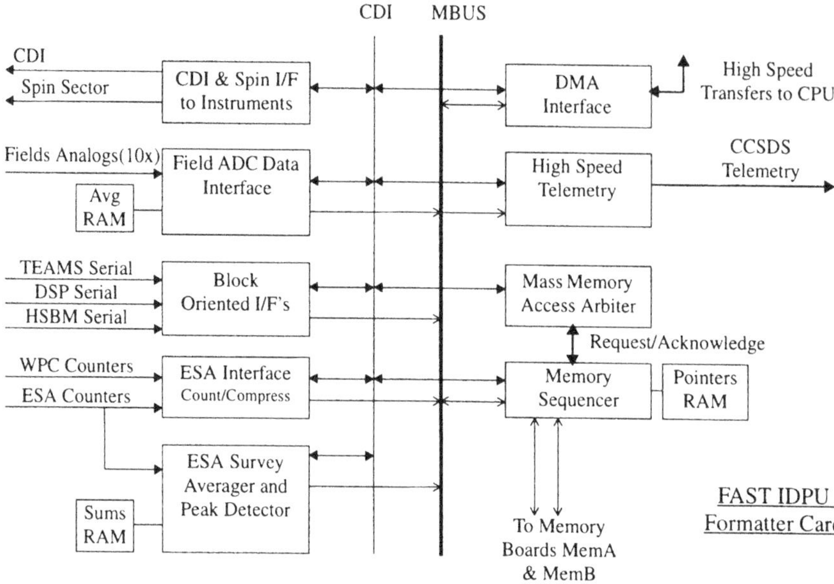

Figure 3. Block diagram of the serial interfaces and functions performed by the FAST formatter board.

nation pointer into memory, and the data is transferred directly to memory as it is collected and formatted. The Processor provides new memory pointers once per frame to each data source, upon request from the Formatter. Access to the memory is arbitrated between the various sources using a custom priority scheme designed to guarantee sufficient memory bandwidth to each source.

The Formatter has two data channels which transfer data from Memory to the high-rate telemetry interface. One is used for CCSDS headers and the other for the corresponding data. The Formatter does very little to most of the collected data before transferring it to memory. Some data from the particle instruments are compressed to 8 bits using a pseudo-square-root compression algorithm (linear for <15 counts, and approximating square-root at larger counts to match the counting statistics). Survey data are generated by the Formatter from the raw data by summing particle data or by averaging fields data. Resolution of the survey data is programmable. In addition, survey data collection involves de-spinning the particle data, since the spacecraft will rotate a significant amount during an accumulation interval. Survey data are collected in parallel with the raw, high-time resolution data to insure complete coverage, since only a fraction of the high-time resolution data are retained for transmission.

The Formatter provides a buffered spacecraft 4 MHz clock (or a lower rate clock divided down from this clock), plus a once-per-second synchronizing pulse to time-synchronous instruments and the Processor. A CDI command is also broadcast to all instruments indicating the Universal Time (UT) at the next once-per-second pulse.

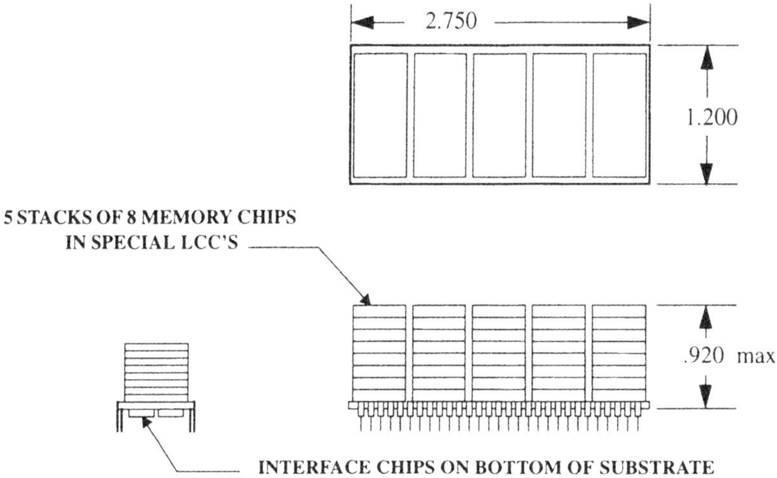

Figure 4. Mechanical drawing of a memory module made by DensePac, Inc., specifically for the FAST mission. Each module is organized as 1 million 40-bit words and weighs about 100 g.

The Formatter provides a sector clock of 12,288 pulses per spin, plus a once-per-spin pulse to spin-synchronous instruments and the Processor. Flight software controls the frequency of this sector clock using a programmable rate generator and phase locks the once-per-spin pulse utilizing spacecraft-provided sun and horizon sensor pulses. The Formatter also calculates a magnetometer spin phase defined as the phase angle between the zero crossing of a magnetometer axis and the sun pulse. Magnetometer spin phase is included in data packet headers to simplify data analysis and is used by particle instruments to control deflectors that track the magnetic field.

2.3. Memory

The FAST Mass Memory system provides the IDPU with 125 MBytes of error-corrected high-speed RAM, distributed on two electrically independent units. The memory uses a stacked-chip memory module made for the University of California, Berkeley (UCB) by DensePac, Inc., which contains 40 Hitachi 128kbits × 8 SRAM memory chips arranged in 5 stacks of 8 chips each, along with all necessary decoders and data buffers. Each module is organized as 1 mega-word by 40 bits. Module packages are about 1.25″ × 2.75″ × 1.0″ in size and weigh very close to 100 grams (see Figure 4).

Each 10″ × 10″ memory board can accommodate sixteen modules, address decoders and two FPGA chips which provide the interface to the board connector and perform single bit error correction on the stored data. The overall organization of each card is 16 777 216 words of 32 bits, with the 40-bit width of the modules providing the additional storage needed for the Error Detection and Correction (EDAC) logic. Since the modules are relatively heavy, each module is glued into an aluminum frame that is bolted to a board-mounted, waffle-shaped support frame. The frame prevents flexing of the circuit board during vibration which could easily break pins on the large memory module packages.

Concern that FAST might experience levels of radiation higher than expected and lose its entire memory system to radiation damage prompted the inclusion of a minimum amount of rad-hard memory. The highest addressed DensePac module was omitted from one memory card and a set of sockets installed instead. This allowed a separate $10'' \times 10''$ card to be electrically connected in place of the DensePac module. The card contains 40 IBM 32 kbits × 8 megarad hard SRAMs, organized as 262 144 words of 40 bits. It has $\frac{1}{4}$ the capacity of the DensePac module. In summary, the FAST memory consists of two memory boards, one of them having the full complement of 16 DensePac modules and the other equipped with 15 DensePac modules and the 'rad-hard IBM board'. This yields a total capacity of 125 MBytes.

Each memory card uses its EDAC circuit to 'scrub' out single-event upsets in the memory by simply reading, correcting and re-writing each memory location periodically. This keeps correctable single-bit errors from becoming multiple-bit (and therefore uncorrectable) errors. Upon command from the Processor, the memory card automatically scrubs a section of its memory specified by a 10-bit block address. The scrubber hardware steps through the identified section of memory, scrubbing each cell and counting both single-bit and multiple-bit errors. When the block is finished, flight processor reads back the number of errors encountered in that memory section and provides these data to housekeeping packets.

A special power control circuit protects the memory modules against latch-up induced over current, and isolates the system against failure of the memory chips on each module. This function, implemented on a FPGA chip, along with analog circuits contained on five small daughter boards mounted on the main memory board, provides a separately controllable circuit breaker on the power line of each module. Each module also contains a set of radiation hard 54AC245 drivers which isolate the system bus from failures in the memory chips.

Distribution of latch-up protected power to the 16 modules is performed with three 16-bit wide control registers called Power, Trip and Override. In normal operation, the Power register is configured to power the memory, turning on a 10 mA 'standby' current source which delivers +5 V to each DensePac memory module. The Trip and Override registers are disabled in normal operation. When a module is 'chip selected', a 300 mA 'active' current source is enabled to supply the extra current needed for operating the memory. If the voltage on any module drops below 4.5 V (in the event of an SEU), the corresponding bit in the Power register is reset and the 10 mA 'standby' current source to the module is turned off.

If the bit in the Trip register is set, the current source is held 'on' even in the event of an under voltage condition. If the bit in the Override register is set, the 300 mA current source is held 'on' continuously. This option would be used in the case of radiation damage to a module which still works, but draws more than

10 mA of standby current. Note that 'chip selecting' a module will always enable the 300 mA source, so to fully shut down a module the software must not address it.

2.4. Power Converter/Switches

Science requirements on the FAST payload specified that data be collected at a high rate during a relatively small portion of the orbit (during auroral crossings), and at a much lower rate during the remainder of the orbit. The spacecraft required operation of different complements of instruments depending upon which data were to be emphasized. The capability of the SMEX launch vehicle limited the size of the solar array and batteries, and drove the payload power system toward a design with a large dynamic range. The power system design required high efficiency both at high power outputs, when all science instruments are collecting data at the maximum rate, and at low power outputs, during the majority of the orbit when limited data are collected. It also had to meet the usual requirements of isolation between power and signal grounds, and have the ability to provide regulated power at various positive and negative voltages to the science instruments while operating from an unregulated 28 $\pm$7 V bus. In addition, to minimize power consumption while taking data with various combinations of instruments, it was required to be able to separately switch on and off the power to each instrument. Finally, to properly manage the system and minimize overall power, the current from each separately switched service had to be monitored.

Figure 5 shows the block diagram of the system used on the FAST payload to meet the above requirements. Two MUE-controlled 28 Volt services supply power to three separate DC-DC converters containing both isolated and regulated outputs. A third MUE-controlled 28 V service is used for boom deployments. Converter number 1 is always powered and supplies power to the IDPU and the instruments which are usually powered during the low data rate portion of the orbit. (This includes the mass spectrometer high voltage, the fluxgate magnetometer, and part of the electric field instrument.) Converter number 2 provides regulated isolated 28 V to unregulated DC-DC converters in the electrostatic analyzer (ESA) particle detectors and electric field boom units. Converter number 3 provides regulated $\pm$12 V and $\pm$5 V to the parts of the electric field analog signal processing electronics which are usually operated only during periods of rapid data taking.

Following the three DC-DC converters is a power controller switch bank which consists of 48 separately limited and monitored power services controlled by 26 logic signals which separately limit and control power to the various parts of the science instrument. A logic signal switches through four different power configurations depending on the requirements of the science instrument being controlled. Current limiters for the higher power services 'fold back' (exceeding the current limit will cause the current to fold back to a low current limit effectively shutting off power) to limit power dissipation in the pass transistors. The lower power ser-

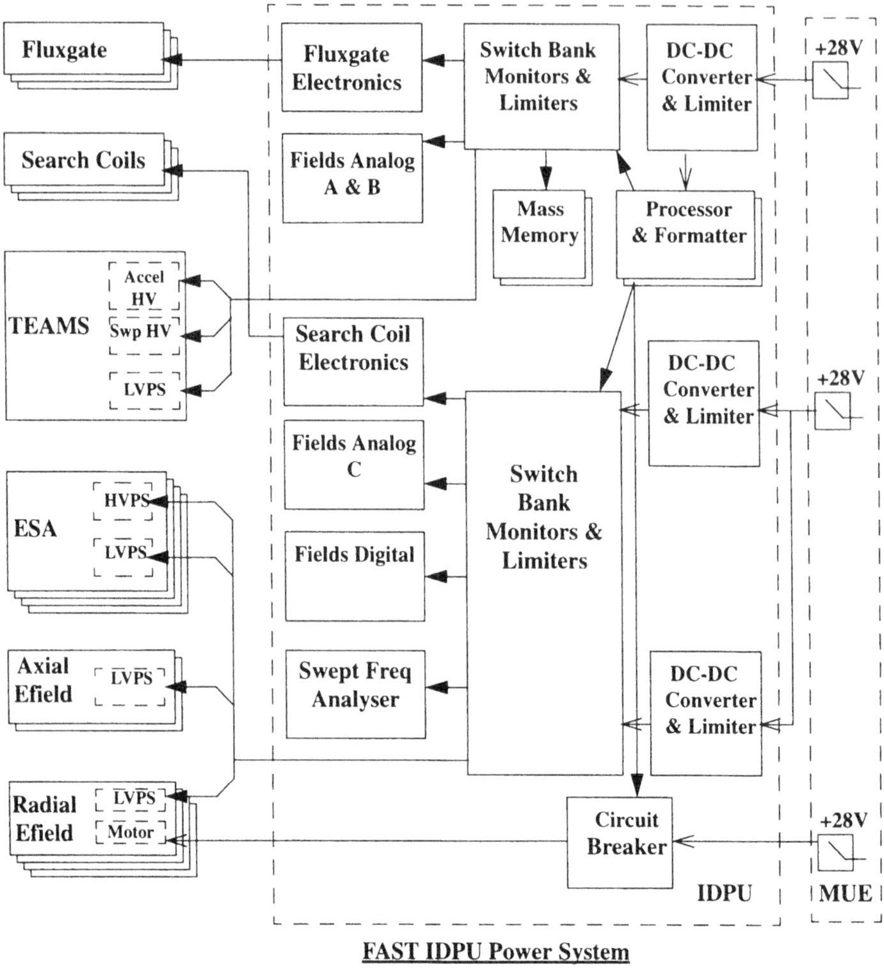

Figure 5. Block diagram of the FAST power system.

vices have a constant limiting current. Both current and voltage are monitored on all services and these values are read by the processor and stored for instrument housekeeping.

The switches to the various instrument services are controlled by the Processor which applies power only to those instruments that are programmed to collect data, write data into the mass memory, or put data into the telemetry downlink. Power can be further reduced by using time tagged commands to request the MUE to turn off the input 28 V to converters 2 and 3 when in low power mode. In this way both high power and low power operation can maintain high efficiency.

3. Software Description

The FAST Instrument Data Processing Unit (IDPU) flight software is responsible for instrument power control, time and attitude determination, mass memory control, science instrument control, command distribution, telemetry formatting and boom deployments. Relying upon the MUE for mission critical operations allows the IDPU software to focus the processing power of the 10 MIPS 32-bit CPU primarily on science data collection and related engineering.

The IDPU flight software is organized into 21 modules, totalling just over 20 000 lines of assembly code. It requires a mere 31 KBytes of code space and 60 KBytes of RAM. The software was written in assembly language and represents about one man year of effort. For radiation tolerance and extensibility, a bootstrap version of the flight code is stored in a bi-polar PROM and later versions are kept in an EEPROM. Upon initialization, the bootstrap software copies itself into RAM, turns off the PROM, and checks over the four available EEPROM programs. Unless commanded to stop within ten seconds, it automatically loads and executes the latest flight software version. Thus, the flight code runs entirely in RAM, requires no PROM or EEPROM power, and is directly patchable by ground command.

The flight software can be functionally grouped into these main elements: (1) Operating System, (2) Engineering, (3) Commands, (4) Telemetry Management, and (5) Instrument Control.

3.1. Operating System

The IDPU includes a very simple operating system for loading and executing new software, handling interrupts, distributing processor time to a variety of tasks, and functionally isolating hardware from software. Included in this functional grouping are three modules which manage the CPU's foreground, background and input/output, respectively.

3.1.1. *Foreground Management*

IDPU foreground management is a simple 'round-robin' polling loop in which each module gets called once per cycle. It is intended for non-time-critical applications and operations that require more CPU time than is allowed in an interrupt service routine. Since all available CPU cycles are expended in this loop, the loop rate is telemetered as a rough gauge of CPU load, and typically runs around 2 KHz.

3.1.2. *Background Management*

The background module is the central clearinghouse for interrupts, containing the logic for controlling CPU interrupts while calling subordinate modules as needed to service each one. Thus, all other modules are isolated from the details of the CPU interrupt hardware, and the background manager is able to level the load. During a typical science data collection, flight software must handle approximately 4000 interrupts per second, divided among five varieties: UART, CLOCK, SECTOR, FORMATTER, and DMA.

A Universal Asynchronous Receiver Transmitter (UART) interrupt occurs whenever a byte is received by or transferred to the MUE over one of the two redundant UART circuits. Under normal conditions, data flows on only one UART and in only one direction at a time, and so the nominal interrupt rate is 960 Hz. The Communication module is called to service these interrupts and determine which or both of the UARTs to enable (see Communications).

The CLOCK interrupt occurs at 256 Hz, driven off the MUE-provided 2^{22} Hz clock line, and in sync with the 1-s ticks. Time keeping, command processing, telemetry buffer management, time-tagging of the sun and horizon sensor inputs, etc., are time-sliced activities performed in this interrupt. This is implemented using a fixed table of vectors so each function gets a known execution rate. For example, commands are called at 128 Hz, the telemetry manager at 64 Hz, and so forth.

The SECTOR interrupt occurs 1024 times per spin, and under nominal 5-s spin conditions, these occur at 1024/5 s or 205 Hz. Under this interrupt, flight software manages the phase-locked-loop control, housekeeping sampling, particle instrument sectoring, and deployment logic.

The FORMATTER interrupt occurs whenever the formatter board completes any of 31 possible data transfers to or from the mass memory system. There are 29 instrument data sources producing 1 KByte packets which the formatter stores into the Mass Memory, and 2 telemetry packets (header and data) which the formatter transfers from Mass Memory to the MUE. On each interrupt, the software determines which memory transfers have completed and calculates the new memory positions for each pointer for the next memory transfers. For each completed data packet, the software formats a CCSDS header block, puts it in a First In First Out (FIFO) memory and sends a request to the DMA module to transfer it to the mass memory. With telemetry on and the IDPU running a typical mode, the formatter card is transferring data at 10 Mbit s^{-1} systemwide resulting in approximately 1600 formatter interrupts per second. To support this rate, flight software is formatting 320 Kbps of header data to tag the hardware-generated telemetry data.

The DMA interrupt occurs whenever packets are transferred between the CPU memory and the Mass Memory. Science frame headers, spacecraft engineering frames, scrubber packets, trigger packets and engineering packets are written to memory while magnetometer data are read out of memory using this facility. The transfer requests are buffered in a software FIFO in order to level the transfer load.

3.2. ENGINEERING

3.2.1. *The Communications Module*

For command and control functions, the IDPU communicates with the Spacecraft Mission Unique Electronics (MUE) using redundant 9600 baud UARTs and a bi-level control line called UARTSEL. After initialization, IDPU flight software receives communicated packets from the MUE from either UART and responds to the MUE using the same UART. Since the MUE automatically switches UARTs in the event of a repeated communication malfunction, this logic keeps the IDPU software in sync automatically. To prevent an errant UART from inventing its own characters and disturbing the communications, the UARTSEL line allows flight operations personnel to force communications to proceed using a selected UART. Software monitors the UARTSEL line and counts transitions to select which UART to use.

The relationship between the MUE and IDPU is strictly master-slave. The MUE initiates all transfers and the IDPU responds, sometimes merely with the equivalent of 'OK'. Each packet from the MUE to the IDPU begins with an opcode and a length byte, and ends with a 20 ms gap. IDPU software checks the validity of the opcode and verifies that the length byte matches the number of transmitted bytes in the packet. In the event of an unknown opcode or mis-match in the length, the IDPU responds with an error code.

Except for ground command packets which occur asynchronously, the communication schedule is defined by tables inside the MUE. Time and backup ACS data are sent to the IDPU once per second, and a status request is performed at the same rate. MUE housekeeping packets are sent to the IDPU at a very low $\frac{1}{8}$ Hz rate. In addition, one of the status bits of the IDPU is a request for more telemetry on a 'space-available' priority.

Across this interface, the MUE sends two types of engineering frames to the IDPU for downlink. Virtual Channel 0 (VC0) frames are real-time engineering frames which the IDPU transmits at high priority while VC1 engineering frames are stored in an IDPU memory segment until commanded to dump.

3.2.2. *Housekeeping Data Collection*

One of the tasks of the flight software is the collection and transmission of housekeeping analog and digital data. Since the A/D converter is power strobed with the 1024 spin sector pulses, the housekeeping module runs at that rate (205 Hz). On each interrupt, the software is allowed to make three A/D samples, which are allocated to 'slow', 'fast' and 'ESA' housekeeping. Slow housekeeping provides a multiplexed slow sampling for all instrument voltages, currents, and temperatures. Fast housekeeping samples a small number of electric field signals or a boom deployment at a higher rate. ESA housekeeping is used to monitor high voltage turn-on. A foreground task monitors housekeeping levels for limit violations and can turn off systems that exceed these limits.

Since the IDPU is slaved to the MUE on the communication port, IDPU software formats CCSDS housekeeping packets only when called to do so by the MUE. This provides the most-recent data to telemetry, but has the disadvantage that the sampling and telemetry are asynchronous.

3.2.3. *Power Management*

In order to conserve power, almost all of the IDPU instrumentation is powered off except while taking data over the north or south auroral zones. The background module keeps instruments informed of which instruments are 'on' by broadcasting the 32-bit power configuration once per second. This information is primarily used by the various electronics boards to disable inter-instrument interfaces in which one of the components is unpowered.

As a safety measure in case the main bus voltage is low, the spacecraft MUE can command the IDPU to go to one of four power levels (0, 1, 2, 3). This is implemented by masking the 32-bit Power Request register with one of four masks, each corresponding to a level. As the selected power level changes, the one-second broadcasts keep instruments aware of the power configuration. At reset, the IDPU is required to be at power level 0, which allows only the CPU and Memory to be powered. Nominal operation leaves the IDPU at power level 3. Intermediate power levels are implemented by the MUE when the battery drops to low charge states.

3.2.4. *Time Management*

The background module uses the CLOCK interrupt to maintain Universal Time (UT) and to ensure that all data generator circuits have the correct time. The MUE regularly sends its Mission Elapsed Time register and a UT offset value to the IDPU, which converts these into a UT clock, formatted as 4 bytes of seconds and 2 bytes of sub-seconds. IDPU software adds 1 second to determine the time at the next tick mark, saves this in a temporary register, and broadcasts this time via the CDI bus to all data generator circuits. When the 1-second tick mark is observed, the temporary register is latched into the IDPU Time register by the interrupt software, while all data generators latch their time registers. The module then uses the 256 Hz interrupt to increment the Time register for use by the rest of the flight software.

3.2.5. *Attitude Determination and Spin Control*

One of the IDPU software functions is to electronically sector or 'point' instrument data in a fixed, sun-referenced coordinate system. If the vehicle is in sunlight, this is pretty straightforward through the use of a sun-sensor and phase-locked loop circuit. However, when the spacecraft enters the shadow, it spins up due to thermal contraction of the spacecraft booms, so maintaining a sun-referenced system throughout the shadow is non-trivial. Since FAST also contains horizon sensors, knowledge of the Earth's horizons can be combined with orbit knowledge to determine the sun-referenced spin phase. Determination of the sun-referenced spin phase is divided into three parts: object timing, spin phase control, and shadow operations.

3.2.5.1. *Object Timing.* Objects are any physical measurement that can be used to determine spacecraft spin phase. Object timing is performed by a combination of Formatter board circuits and software which puts UT time-tags on MUE-supplied Sun and Horizon-crossing signals. Redundant time-tagged versions of these are also communicated via the IDPU-MUE communications link. Both hardware- and MUE-reported Horizon crossing signals are used to calculate the Nadir spin phase. This calculation is slightly complicated by the fact that the Sun and the Moon also trigger the Horizon sensor, but these objects are easily identified by their relatively small angular widths. Sunrise and sunset cause a unique signature in the horizon pulses that is easily handled by the Nadir calculation. IDPU software constantly maintains and telemeters the sun crossing times and earth nadir crossing times from both primary and backup sensors. Magnetic field sensor zero crossings are used to calculate an additional object time.

One of the above object times is selected for spin phase determination and flight software maintains a virtual object which is a copy of the selected object's time plus an optional offset. A bit in the object selection register enables a time-offset register to be applied to the object time, so that sectoring will point ahead or behind a true object. This feature is used for correcting the differences between the response of sensors and to allow Sun pointing to be calculated from Nadir pointing, as described later. Finally, software calculates the object's period using an infinite response digital filter, rejecting invalid points prior to filtering. Points are deemed invalid when too large a phase shift is observed.

3.2.5.2. *Spin Phase Control.* Spin phase control is maintained by a combination of hardware and software. Using a programmable divider of the 2^{22} Hz clock, IDPU hardware broadcasts a 'spin-sync' pulse and 12×1024 'sector' pulses per spin to all instruments. Phase synchronization between 'spin-sync' and the selected 'object' is maintained by a software 'phase-locked loop' that adjusts the clock divider value. The allowable range in hardware is 512 to 4096 which corresponds to spin periods of 1.5 to 12 s.

Since a nominal spin period of 5 s results in a divider value of 1280, maintaining a fixed value over a spin would yield only a modest 0.3 deg precision. Thus, flight software calculates an interpolation point within the 1024 sector spin in which to switch this divider from 'N' to '$N + 1$', giving a much more precise solution. For example, if the spin period is 5.000015 s, the divider should be 1280.0039. This is approximated by $0.0039 \times 1024 = 4$ sectors at 1281 and 1020 sectors at 1280.

Software maintains phase synchronization by comparing the times of the local spin-sync pulse and the selected object. When this error is above 1 degree, it calculates the divider value that will achieve an immediate match between spin-sync and

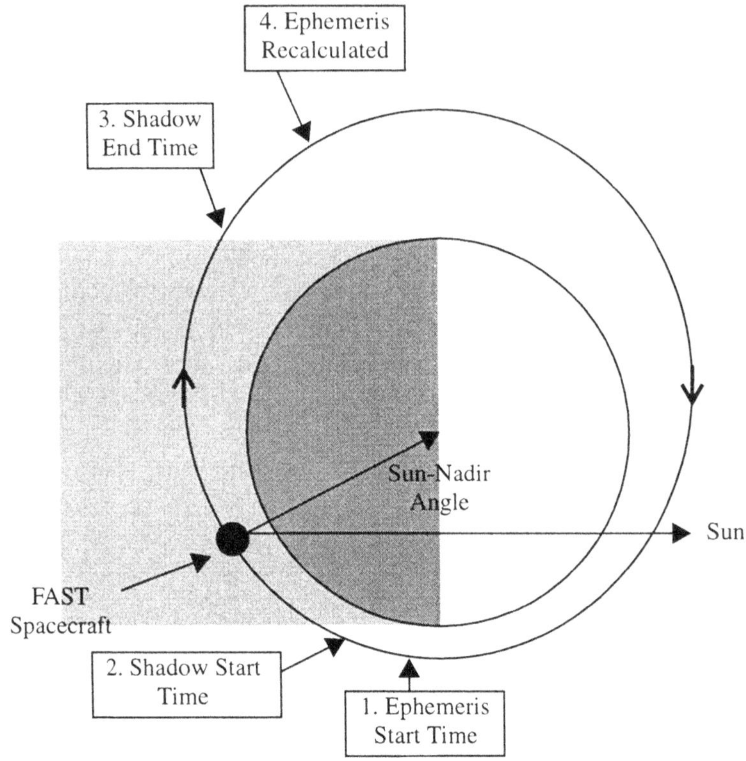

Sun-Nadir Operation Points of Interest

Figure 6. Drawing showing a sequence of time-tagged events that are used to turn on and off the Sun-Nadir routine. The Sun-Nadir routine allows the spacecraft to calculate the sun direction while in shadow using knowledge of the nadir direction provided by horizon sensors. This routine provides for smooth transitions in and out of eclipse for data organized by spin phase.

the object using the known period. For small error-angles, software uses a filtered phase error to determine the divider value and when to flip from 'N' to '$N + 1$'.

3.2.5.3. *Shadow Operations.* Shadow operations are implemented to provide smooth transitions in and out of eclipse. The role of the 'shadow' software is to keep the sectoring pointed at the sun, even during shadows. To do this, the Science Operations Center or SOC (see McFadden *et al.*, 2001) calculates a Sun-Nadir ephemeris file which is loaded into the IDPU once every few days. Included in this file are the orbit period, the UT to start using the new file, the shadow start and end times, and two tables. The first table is a piece-wise linear approximation of the Sun-Nadir angles during the first orbit, and the second is the differential orbit-to-orbit change in Sun-Nadir angles. The two tables can be combined to generate Sun-Nadir angles that are accurate enough to operate for several days.

As shown in Figure 6, the shadow sequence begins at an Ephemeris Start Time. Using the piecewise linear angles array, software begins calculating the offset between the earth and the sun once per second. The start time is in sunlight to allow comparisons between the real and the calculated Sun-Nadir offset. The pointing object is still the sun, and the sun-nadir offset is only telemetered. At the Shadow Start Time defined in the ephemeris, software switches the selected object to 'Nadir-plus-offset', and calculates the difference between the true and calculated sun angle to generate the 'onset error'. The 'onset error' is used to force agreement between the ephemeris-determined sun crossing time and actual Sun crossing time. Thereafter, software simply calculates the Sun-Nadir angle, subtracts the onset error, and sets the object offset. (Note: This has been implemented so that UT does not need to be monotonic. Spacecraft UT can be adjusted without problem.) Finally, at the Shadow End Time, software switches the object back to Sun, and recomputes the ephemeris for the next orbit. The Ephemeris Start Time is updated by adding the orbital period, and the delta-angle table is simply added to the angle table. The ephemeris gradually gets worse at predicting the position of the Sun, and computer models show a maximum usable duration of about a week. On-orbit performance has been superb, routinely demonstrating better than 1 deg pointing accuracy in shadow.

3.2.6. *Deployments*

Flight software responsibilities include deploying the four spin-plane wire boom systems, and verifying the deployments of the axial and magnetometer booms. Each boom unit is equipped with a turns-counter microswitch which is sampled by the IDPU software and converted to a length accurate to about 10 cm. Software is normally commanded to deploy a pair of booms on opposite sides of the vehicle to a length defined in 'turns' or 'clicks' on a microswitch closure caused by a drive-wheel cam.

Since booms deploy at slightly different rates, software monitors the lengths and if one boom gets more than 2 'clicks' ahead of the opposite boom, the longer boom is paused until the shorter boom catches up. If either boom indicates a 'jam' condition, the deployment is halted within 0.1 s.

A unique feature of the FAST deployment software is that it synchronizes deployment with the oscillation period of the deployed booms in order to cancel boom motion. Decisions to start and stop motors are made in phase with the predicted oscillation, so the booms are not swinging at the end of each deployment. On-orbit performance of this logic has been verified.

3.2.7. *Mass Memory Control*

3.2.7.1. *Memory Integrity.* In order to maintain memory integrity in the presence of penetrating radiation, the memory has an Error Detection And Correction circuit (EDAC). This circuit is capable of correcting all single-bit errors and some multi-bit errors during the read cycle. Using the EDAC, a 'scrubber' circuit simply reads

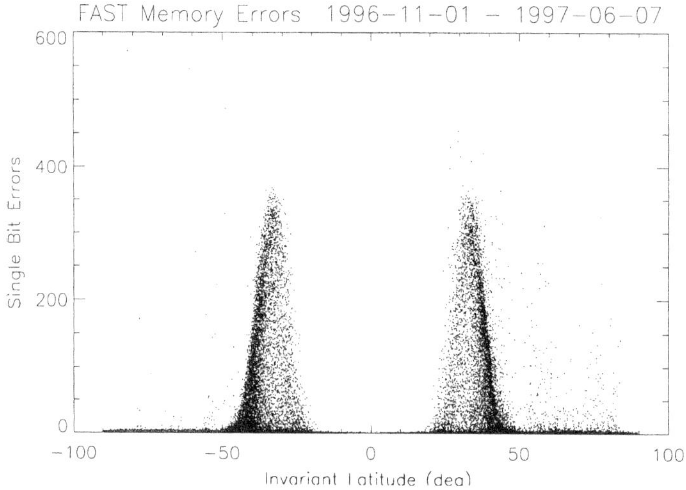

Figure 7. The graph shows the invariant latitude location of single bit errors in the mass memory. Most errors are associated with penetrating radiation as the spacecraft passes through the inner radiation belt. Single bit errors are corrected by Error Detection and Correction (EDAC) hardware and scrubber software which periodically pages through the memory.

and re-writes memory locations correcting single-bit errors before they become uncorrectable double bit errors.

The flight software initiates scrubs on each of the 1024 × 128KB pages using an 8-bit timebase, whose default is 4 pages per second. On each scrub cycle, the software reads the number of multiple and single-bit errors found in the page, and generates 'scrubber' telemetry packets. The software keeps the total number of errors for the entire valid memory, as well as identifying the worst memory segment with its error count. These data are available in the housekeeping telemetry.

On-orbit performance of the scrubber circuit has demonstrated that the corrections are usually single-bit, correctable, and uniformly distributed throughout the memory. In a seven month study (1447 orbits), the errors were found to occur primarily between ± 50 deg of invariant latitude, within the radiation belts, as seen in Table I and Figure 7.

3.2.7.2. *Memory Power.* As described in the memory hardware section, the memory subsystem is powered in 32 sub-sections, each with its own current trip circuit. This strategy guards against the possibility of a single latchup disabling the entire memory. The software monitors the status of these 32 trip circuits and includes it in the memory scrubber packet telemetry. Only one memory trip was observed during the first two years on orbit. This module was reset and continued to function properly.

3.2.7.3. *Virtual Memory.* Concern that parts of the Mass Memory might fail either during launch or due to radiation prompted IDPU software to provide a

TABLE I

During a seven month study, memory errors occured primarily within the radiation belts

Item	Single-bit errors	Multiple-bit errors
Total error count	918383	2915
Errors per orbit (7990 s)	630 ($\pm$320)	2 ($\pm$1.8)
Errors | ILAT | < 50 deg	96.8%	93.7%
Errors | LAT | < 50 deg	97.2%	94.9%

virtual-to-physical mapping of the memory. The software remaps the upper 10-bits of the address in order to subdivide the memory into 1024 pages of 128 KBytes each. Using a bitmap loaded by ground command, the memory controller software recalculates the virtual-to-physical mapping automatically, and adjusts the amount of telemetry allocated to each Virtual Channel. Thus, bad sections of memory can be removed from operation, while the only observable effect is that the Mass Memory shrinks! After 2 years of operation, only one page of memory has been removed because it contained unusually high bit errors.

3.2.7.4. *Memory Configurations.* The Mass Memory is used exclusively for CCSDS frames, each being 1064 bytes in length. In order to make the hardware processing of science data easier, the 40-byte header and 1024-byte data sections are stored in separate areas of the Mass Memory. All science data is provided by the Formatter directly into the 1024-byte sections, while the IDPU formats and writes all headers to the Header Segment.

All 40-byte headers are packed together into the Header Segment, abiding by the hardware restriction that no header cross a 1 K boundary. Thus, only 25 headers are stored per KByte. The size of the Header Segment is determined by how much usable Mass Memory is available. The software calculates this and uses only the amount of memory necessary. If all 125 MBytes of memory are available, the headers require 5.17 MBytes for the 126144 packets.

3.3. Command Management

3.3.1. *Command Strings*

The command module executes strings of 32-bit commands, either sent from the ground through the spacecraft MUE or commanded internally by other IDPU software modules. If no spacecraft commands are present, flight software executes command strings from internal sources. New instrument operating modes, Fast and Slow Survey data rate transitions, and instrument initialization sequences are examples of internal strings of commands.

While commands are nominally executed at a rate of 128 Hz, a delay command is provided for slowing down command sequences as required in some cases. The module also provides a counter function in order to verify reception of long ground-commanded sequences.

3.3.2. *Memory and Table Operations*

The command module includes a very basic set of memory commands for loading, dumping and executing. In addition, there are 18 tables which allow memory operations to use relative addressing. This facility allows loads to be somewhat independent of software version.

A special feature of memory loads involves the EEPROM. Since writing to the EEPROM is limited to 1000 writes, software checks each word to see if the EEPROM already has the value and writes data to it only as necessary. In addition, performing EEPROM writes requires an enable command.

3.3.3. *Modes*

The command module is responsible for loading, checking and executing special sets of command strings known as 'modes'. There are two basic varieties of modes, Particle modes and Field modes. The IDPU stores up to eight of each type, with each mode containing a maximum of 512 commands. Modes are identified using an 8-bit code and periodically verified using an integral checksum.

Modes are executed using a 'start mode' command which requests one Particle and one Field mode identifier. In response, the command module simply looks up these command strings and uses its internal command channel to execute the strings.

It is worth noting that a variety of IDPU operations, particularly instrument driver commands, are implemented by spawning command strings for instrument configuration. When executed from within a mode, these spawned command strings must wait for the mode string to complete in order to execute.

Contained in each mode definition are two command sequences, one for Slow-to-Fast and one for Fast-to-Slow survey data rate transitions. These sequences contain a maximum of 256 commands and define for the IDPU what it should do when switching between different survey data collection speeds. Thus, a given Fast-to-Slow or Slow-to-Fast transition can have up to 512 total commands to tell the science instruments how to change the data rate.

3.4. Telemetry Management

3.4.1. *Virtual Channel Allocations*

In order to handle the various data types (housekeeping, survey data, burst data, etc.) with different readout priorities, the Mass Memory is partitioned into several blocks. Flight software subdivides the Mass Memory into eight Virtual Channel storage areas, an area for packet headers, a VC7 (fill) packet, and a spare packet

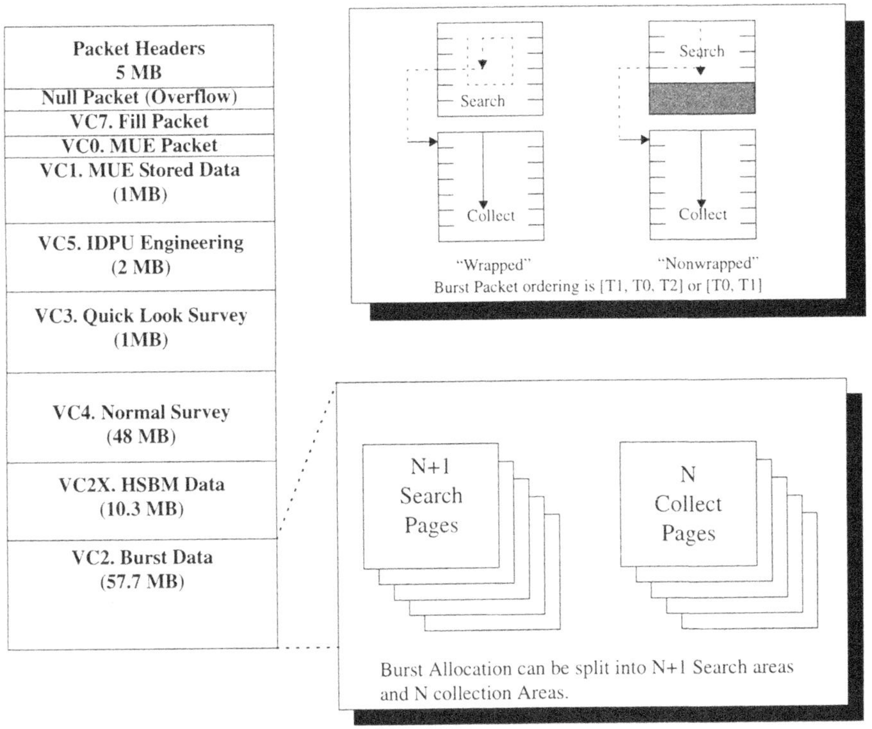

FAST IDPU Mass Memory Allocations

Figure 8. Diagram of the FAST mass memory allocation into Virtual Channels (VCs). Most memory segments or VCs are operated as FIFOs. VC2 burst data differs as shown with this area of memory further subdivided into N collect and $N + 1$ search pages. One collect and one search page combine to form a single burst collection, with readout priority given to the burst with the highest quality. The extra search page is used to temporarily store data until a trigger algorithm selects the new data and a collect page of the least interesting burst is overwritten.

to handle full-memory cases. Figure 8 illustrates these storage areas and Table II shows typical size and allocation priorities. During a memory reconfiguration, memory is allocated on a table-driven priority basis, with 0 the highest priority. Lower priority virtual channels may not get their full allocation if the sum of the requested memory is less than the available memory, which may occur if memory modules are disabled.

3.4.2. *MUE Engineering Telemetry Management (VC1)*

The VC1 segment is used to store MUE Engineering telemetry frames sent to the IDPU via the IDPU-MUE interface. These MUE generated packets contain housekeeping data on various spacecraft systems (command status, battery state of charge, temperature monitors, voltage/ current monitors, attitude control inputs, etc.). MUE telemetry frames are stored in a circular fashion, with the newest data

TABLE II

Typical mass memory allocation among various data products including memory allocation priority with 0 the highest priority

Segment	Allocation	Priority	Description
HEADERS	5.2 MB	0	Header area for all frames
NULL	0.001 MB	0	Location to direct overflow packets
VC7	0.001 MB	1	CCSDS fill data frame
VC0	0.001 MB	2	MUE real-time engineering frame
VC1	1.0 MB	3	MUE stored engineering data
VC5	2.0 MB	4	IDPU engineering data
VC3	1.0 MB	5	Survey data (quick-look)
VC4	48.0 MB	6	Survey data
VC2X	10.3 MB	7	High speed burst memory data
VC2	57.5 MB	8	Burst data
Total	125.003 MB		

overwriting the oldest. A playback command dumps the contents of the buffer, but does not automatically clear it. Thus, the flight controllers can repeat the playback without losing any data. A separate 'clear' command resets the buffer.

3.4.3. *IDPU Engineering Telemetry Management (VC5)*

Similar to the VC1 segment, the VC5 segment is a circular buffer used to store IDPU Engineering frames generated by IDPU flight software. These blocks are stored in the circular FIFO until the transmitter is available. In the event of overflow, the IDPU saves the most recent engineering data by overwriting the oldest data.

3.4.4. *Survey and Quick-Look Telemetry Management (VC4 and VC3)*

In order to provide a large-scale context for the high-rate burst data snapshots (see Section 3.4.5), a 'survey data' collection scheme was implemented. Survey data are generated by all the instruments during science data collection, either by averaging the data or by filtering the output of analog signals before digitization. Two different survey data collection rates, termed fast survey and slow survey, are used to optimize the science. Fast survey data collection rates ($\sim$0.5 Mbit s^{-1}) are about an order of magnitude higher than slow survey rates ($<$0.05 Mbit s^{-1}) and about an order of magnitude lower than high-rate burst data collection ($\sim$5–12 Mbit s^{-1}, see Section 3.4.5.). Onboard data evaluation schemes determine which survey collection rate to use and a memory management scheme prevents survey memory overflow (described below). In addition, survey data can be stored in two separate memory buffers (VC3 and VC4) which have different readout priorities and are used to enhance real-time science operations.

The VC3 and VC4 segments are independent circular FIFO buffers providing comprehensive storage of survey science data. While VC4 provides the bulk of the storage, VC3 is intended for quick-look datasets during telemetry contacts. By simply directing a subset of the survey data to an empty VC3 buffer, users can avoid waiting for VC4 to empty before getting real-time survey data.

In addition to storing and playing survey data, the VC4 manager is responsible for switching the IDPU between Fast and Slow survey modes. The concept is simple: whenever the trigger value 'Flevel' exceeds the commanded threshold, a transition to Fast survey data collection mode is initiated. To prevent memory overflow and trigger oscillations, a memory management scheme was implemented as outlined below.

First, in order to prevent the system from simply filling up all of the survey buffers with a few minutes of Fast survey data, leaving no room to complete the orbit, the VC4 manager limits Fast survey data collection to a commandable 'FSALLOC' page limit. The page limit is set by stored command after each data dump, with page limit allocations calculated by the science operations center based upon contact schedules and data collection periods (see McFadden *et al.*, 2001). The Fast survey page limit is implemented by maintaining a bitmap of the VC4 memory, indicating which pages are Fast Survey data and which are Slow Survey data, and keeping track of how much Fast survey is in memory at a given time as data is collected and telemetered.

Second, to prevent system oscillation between Fast and Slow survey data collection during contacts where the data is being simultaneously collected and transmitted, software limits the transition rate. These oscillations would happen when the Fast survey data collection rate exceeds the telemetry rate, and the Slow survey collection rate is below the telemetry rate. For example, the system would go into Fast mode, filling the VC4 memory until it hits the FSALLOC limit, switch to Slow mode allowing the telemetry to catch up, then quickly switch back to Fast mode. To limit this switching, the software allows Slow-to-Fast transitions only if there is $\frac{1}{8}$th of the FSALLOC memory available for Fast survey data.

Third, in order to prevent oscillations between Fast and Slow modes caused by fluctuations in the Flevel trigger calculations, the software only allows transitions to Slow mode when the Flevel falls below the threshold for 20 s. Hysteresis in this decision could have been employed, except the performance would have depended on filtering characteristics, which was judged more difficult than simple timing.

Once the software has decided to switch between Fast and Slow modes, the actual implementation is quite easy. The Slow-to-Fast and Fast-to-Slow transitions are simply command strings set in memory and maintained by 'mode' definitions (see Section 3.3.3), so the VC4 manager merely instructs the command module to execute one of the strings.

3.4.5. *Burst Telemetry Management (VC2)*

The most complex of the storage areas is VC2 memory. It can operate as either a circular FIFO like VC3 and VC4, or as a collection of smaller 'burst' memories. When operated in 'burst' mode, the memory is subdivided into $N + 1$ Search areas and N Collect areas, where N is up to 64. Search areas are used to store data before a triggered event and Collect areas contain data sampled after the event. For each burst collection, there is a 16-bit 'science merit' evaluation of the data within that collection called EVALMAX. This value is calculated by a weighted average of the Burst 'Goodness' or 'Bgood' value during the Search and Collect periods. 'Bgood' is evaluated from trigger inputs, with an algorithm similar to the trigger algorithm as described below in Section 3.4.6.

To start a burst collection, the flight software finds the open Search area and directs all VC2 data sampling into that area of memory. At 10 Hz, the Burst trigger level or 'Blevel' parameter is calculated and compared to the minimum of the EVALMAX values for all other stored bursts which are not currently being transmitted. If the 'Blevel' seen during the Search period exceeds the EVALMAX value of another burst, then that burst is selected to be overwritten. Flight software directs all new VC2 data into the overwritten burst's Collect area of memory. When the Collection area is filled, the EVALMAX parameter is calculated and the process begins again.

Flight software selects bursts to be played out on a 'best-first' policy, and, for simplification, does not retract a decision once made. Thus, it is possible that N great events show up during the playing of a less interesting event, and that only $N - 1$ can be stored. Given a large enough 'N', however, this should be very unlikely. FAST typically operates with $N = 5$ which gives bursts with about 10 s durations.

3.4.6. *Trigger Calculations*

Among other things, the IDPU software has to distinguish interesting data from uninteresting data. This is accomplished by applying a pair of user-selected algorithms, called BSALG and FSALG, which control Burst collections and Fast Survey mode, respectively. The user selects which two functions to use, consistent with the planned mode of the instrument, from a library of available functions. These include eight electric field inputs; four electric field functions that combine the electric field inputs to trigger on AKR, Electrostatic Shocks, VLF, etc.; and three ESA functions to trigger on electron and ion counting rates or on a function that tracks the energy with the highest counting rate.

Software executes BSALG and FSALG every 64th of a spin (about 13 times per second), and filters the outputs using three slope-sensitive 16-bit infinite response digital filters, producing the quantities BGood, BLevel, and FLevel. Each filter responds to rising signals with different parameters from falling signals, allowing users to make filters that have both good sensitivity and long retention. BGood is a longer averaged version of BLevel, and the two are used by the Burst manager in the determination of burst collection quality (see Section 3.4.5). Similarly, Flevel is used by the Survey manager to determine Fast survey data collection (see Section 3.4.4).

In order to avoid false triggers during mode transitions and power switching, the module includes an inhibit timer which is usually set by the start of a mode. While active, this clears Flevel and Blevel so that no Bursts or Fast survey modes are activated. In addition, trigger calculations are disabled during both electric field diagnostic sweeps and during trigger algorithm changes.

The trigger packet provides engineering diagnostic data needed to identify 'interesting' regions. Basically all of the available raw trigger data is provided along with the filtered results of two selected functions. Thus, even if the selected functions do not trigger on interesting regions, the raw data can be used to determine which functions would have triggered.

3.4.7. *Telemetry Playback*

Flight software handles up to 530 interrupts per second in processing telemetry frames, feeding the formatter card with the addresses of frames and headers to be telemetered. Each of the VC managers passes segment addresses and lengths (each segment is 16 to 64 telemetry frames) to the formatter control module for transmission. The formatter module saves the segment addresses in a 2-stage list for each VC, providing quick access to addresses yet allowing VC managers to operate at a relaxed pace.

In order to deal with ground segment limitations, the IDPU has to limit data rates by VC (see Table III). For example, VC4 data is limited to 1 Mbit s^{-1}, VC3 to 50 kbps, VC6 to 100 kbps. Software implements frame-rate limiting using 3 rate tables, one for each telemetry rate. In each table are the allowable rates for each of VC1 through VC2X. (Note: VC2X is treated as a separate virtual channel by the IDPU onboard, though its CCSDS header identifies it as a VC2 during transmission.) As telemetry frames are processed, software keeps 6 running rate-adders whose overflows enable respective VC channels for transmission. Once a VC frame is transmitted, another of the same type cannot be transmitted until its rate-adder overflows again. Naturally, spacecraft real-time engineering data (VC0) is not rate limited.

Of the VC types enabled for transmission, the IDPU selects the next VC by simple ordering: i.e VC0, VC1... VC2X. If none of these frames is available to transmit, it will try VC2 (Burst data) and, if nothing else, use the 'Fill' packet VC7. The net effect of all of this is that priority is given to the low volume engineering data (VC0, VC1 and VC5), then to real-time science data (VC3 and VC4), and then to Burst (VC2) and HSBM (VC2X) data.

Finally, in order to allow the ground station to acquire the carrier and determine good quality before dumping science data, each of the VC's can be enabled or disabled for transmission. Software automatically disables all science VC's whenever

TABLE III
Telemetry rate limits for FAST data channels

Telem. priority	Virtual channel	Data bandwidth at 900 KHz	Data bandwidth at 1500 KHz	Data bandwidth at 2250 KHz
1	VC0	Unlimited	Unlimited	Unlimited
2	VC1	49.2	500	563
3	VC5	Unlimited	Unlimited	Unlimited
4	VC3	49.2	49.2	49.2
5	VC4	900	1000	1000
6	VC2	Unlimited	Unlimited	Unlimited
7	VC2X	98.4	100	200

the science telemetry port is cycled, and the ground station enables science VC's only after a good link is established.

3.5. Instrument Control

The FAST science instrument complement is complex, having hundreds of operating modes, requiring active control of data collection at >10 Mbit s^{-1}, and generating telemetry at several Mbit s^{-1}. In order to deal with this complexity, the flight software includes several methods of instrument control: (1) direct commands using the Command and Data Interface (CDI); (2) Mode strings which organize instrument configurations; and (3) the instrument drivers for dynamic control needs. While the first two have been discussed earlier, the following section sketches three software drivers for handling Fields, ESA and TEAMS control functions.

3.5.1. *Field drivers*

The electric and magnetic field experiments include a number of signal processing and data capture systems that must be actively managed by the Field Drivers (see Ergun *et al.*, 2001, for a description of the experiments). These drivers control data collection by the digital signal processor (DSP) and high speed burst memory (HSBM), monitor the magnetic field, determine the optimal tracking frequency for the swept frequency analyzer (SFA), and control electric field probe voltage and current biases. These drivers play a crucial role in assuring that the electric and magnetic field instruments are operated in proper configurations and that data collection is coordinated to maximize the science return.

3.5.1.1. *Digital Signal Processor.* The Digital Signal Processor (see Ergun *et al.*, 2001) is a 32 MHz AT&T DSP32C, performing Discrete Fourier Transforms (and more) on digital data from the Field analog boards or the High Speed Burst Memory. The DSP driver in the IDPU is very simple, controlling the DSP-to-Mass Memory packet transfers and formatting frame headers with mode information. Subsecond timing data from the DSP is transferred to the processor, through the formatter interface, at the end of each data packet and is included by the driver in the packet header.

3.5.1.2. *High Speed Burst Memory.* The HSBM system (see Ergun *et al.*, 2001) digitizes four analog signals at up to 2 MHz and stores these data in one of several 2.5 Mbyte buffers on the HSBM board. Using internal triggering logic, with 10 available signals to choose from, the HSBM autonomously determines and transmits the 'best' data encountered while its acquire signal (ACQ) is active. The functions of the Fields HSBM driver are to raise the ACQ signal whenever Fast Survey or Burst collections are occurring, and to control the HSBM-to-Mass-Memory packet transfers. Since the processor does not have timing information about the HSBM data, the HSBM internally latches the start time of each buffer and sends this data to the processor, through the formatter interface, for inclusion in the packet headers. Since a buffer requires a large number of frames, only the last frame of an HSBM transfer is marked with this latched time. Finally, the driver enables HSBM transfers only if there is sufficient memory storage in VC2X, so that only whole buffers are sent in telemetry.

3.5.1.3. *Magnetic Field Strength Calculations.* In addition to collecting data packets and formatting headers for the Fluxgate Magnetometer, flight software samples the magnetometer data directly by performing DMA transfers from the Mass Memory to the IDPU local memory, and extracting B1, B2 and B3 from the frame. Each 16-bit magnetometer value has its DC offset removed, is scaled once to reverse the effects of the hardware digital filtering in use, and is scaled a second time to convert to 2.5 nT resolution. From these values, software calculates the unfiltered magnitude from the sum of the squares, as well as a filtered version using a 3-to-1 infinite response filter. The unfiltered and filtered magnitudes are then used by the ESA and SFA drivers, respectively.

3.5.1.4. *Swept Frequency Analyzer (SFA) Tracking.* In order to allow high frequency resolution measurements, such as auroral kilometric radiation (AKR) near the electron cyclotron frequency, a frequency tracking driver was developed. When enabled by ground command, the SFA driver software calculates and sets the ideal tracking frequency for the SFA board using the computed magnitude of the magnetic field. The frequency is calculated as TRK $= (0.026817 \times \mathrm{MAGnT} + 10255.3)$ Hz, where MAGnT is the magnitude of the magnetic field in nanoTesla.

3.5.1.5. *Configurations for Northern and Southern Hemisphere.* Some of the instruments and trigger algorithms require knowledge of the nadir direction relative to the magnetic field, which changes with Earth hemispheres. The fields driver software provides command strings which configure the instruments for Northern and Southern hemisphere data collections. Ground system provided time-tagged commands keep the driver aware of the location of the spacecraft relative to the ecliptic plane, so that when modes are started, they need only request the driver to configure instruments for the appropriate hemisphere.

3.5.1.6. *Sensor Bias Tables.* In order to organize the voltage and current biasing of the ten electric field probes through varying plasma densities, flight software manages four 'bias tables'. Each table is implemented as a command string of up to 31 commands which are loaded into a fixed area of local memory by special field modes. Typically, each bias table holds ten biasing commands, one for each probe, and each command contains two bias values, one for voltage mode and one for density mode. In executing each bias command, software simply checks the voltage/density status of each probe in order to command the appropriate bias value.

3.5.1.7. *Sensor Bias Sweeps.* Flight software includes a bias sweep module for in-flight determination of sphere photo-emission current and other characteristics. The number of steps in a bias sweep is programmable. For operational simplicity, sweeps can be initiated by direct commands (from modes for example) or run automatically on a timebase. Software separates these two types such that 'automatic' and 'commanded' sweeps occur without interfering with each other. For example, a direct command can sweep a different pair from those set for automatic mode.

Automatic sweep timing is set using a 4-bit parameter 'ASWP' which requests sweeps at $2^{\mathrm{ASWP}} \times$ spinperiod/32. For example, an ASWP of 5 requests a sweep every spin. Setting ASWP to 15 disables automatic sweeps. If one sweep is still in progress when another is requested, the new sweep is delayed until the prior one ends. This facilitates continuous sweeping profiles.

In performing a sweep, software first marks the 'Sweep-in-Progress' bit in order to inhibit the trigger module and then records the bias settings of the probes. If requested, software waits for a given magnetic angle and/or one-second boundary to begin sweeping probes. Once these requirements are met, software simply steps a selected sphere (or pair of spheres) from one bias point to another (and optionally back again). When biasing a pair of spheres, the software will optionally sweep them together or alternate a full sweep on sphere A followed by B. Once the bias stepping is finished, software places the original bias values onto the spheres and removes the triggering restriction. The data is collected and telemetered by normal VC4 (survey) real-time sampling.

Bias steps are implemented under interrupt at 128 Hz using unused command channel capacity. The step rate can be slowed in factors of two from 128 Hz to 1/256vHz, or can be synchronized to 1 point per spin.

3.5.1.8. *Automatic Re-configuration for Sunlight and Shadow.* Early in-flight performance of FAST showed that during shadow periods, the electric field probes in the density mode caused the spacecraft potential to become negative, attracting ions into the ESAs. Due to orbit mechanics and a limited stored command volume, it was deemed impractical to use stored commands to switch probe modes to match shadow periods. Instead, this function was added to the flight software.

The Sun–Shadow manager operates at a 1 Hz rate, determining whether or not the vehicle is in sun or shadow by watching sun pulse times and defining a shadow as twelve seconds without a sun pulse. Whenever a transition in or out of the shadow is detected, flight software creates a command set and submits it to the command string processor. Following that, the module rebiases the probes by requesting the appropriate 'sun' or 'shadow' bias table.

3.5.2. *The ESA Driver*

The electrostatic analyzer (ESA) driver actively manages the 16 electrostatic analyzers on the FAST payload. This driver controls low and high voltage turnon and turnoff; performs pulse height distribution tests of the microchannel-plate detectors and noise and electronic tests of the preamplifier-counter circuits; initiates high rate ESA engineering data collection; determines the tracking energies for the 12 stepped electrostatic analyzers (sESAs); calculates the particle energy flux and peak energy for trigger algorithms; controls the ion and electron ESA deflector settings which provide continuous monitoring of the field aligned fluxes; and actively monitors the health of the ESAs. The ESAs are typically operated with two pairs of ESAs sweeping over the entire energy range (3–30 000 eV) and measuring ions (iESA) and electrons (eESA) with 78 ms resolution. The 12 remaining ESAs form the stepped ESA (sESA) data product. The sESA performs high time resolution (1.6 ms) measurements of electrons over a selectable energy range. The ESA driver performs instrument commanding through a small set of transitions sequences (or command sets) outlined below. The driver also programs hardware on the formatter board to perform many of the sESA and trigger calculations.

The ESA driver software has a small number of operational modes as shown in Table IV. Transitions between modes are made by Transition Sequences, started by ground or stored commands. These Transition Sequences consist of lists of operations to be performed to change the driver to a selected mode. Before starting a transition sequence, the software first checks to see that the sequence is legal. For example, the 'Turn High Voltage On' sequence cannot be run unless the high voltage enable key has been set. The sequence execution rate is limited by CDI command bandwidth considerations, and may also be intentionally slowed down with embedded pauses for other timing considerations. Some Transition Sequences are used for instrument tests and calibrations, and return to the original mode at the end of the sequence.

TABLE IV

ESA operational modes controlled by the ESA driver

ID	Mode	Description
00	OFF	ESA power is off and driver is idle
01	Low voltage On	ESA low voltages are on and driver is active, but high voltages are off
02	High voltage On	ESA is fully on
16+n	Transition	In the process of sequencing from one mode to another, using transition sequence number '*n*'

3.5.2.1. *Commanding* The default software operation is based on a table of parameters called the ESA Instrument Configuration Table (ESA_ICT). The ESA_ICT default values are loaded upon processor reset from the IDPU EEPROM, and thereafter can be modified by the IDPU Table Load Command. This table includes high voltage levels, sweep rates, preamplifier gain settings, etc.

In addition, there are a few commands which control the ESA driver. These include the high voltage enable/disable commands, diagnostic telemetry request commands, a high voltage ramp-up command, and a Transition Sequence request command.

3.5.2.2. *Transition Sequences.* There are 8 transition sequences, started by the IESASEQ command. Special cases of this command are the IESALVON, IESAHVON, IESALVOFF, and IESAHVOFF commands, which start sequences 0, 1, 2, and 3, respectively. Sequences 4, 5, and 6 are diagnostic test sequences. Sequence 7 is a spare for later use.

3.5.2.2.1 Sequence 0: IESALVON. IESALVON can only be run from the OFF mode (00), and transitions to the Low Voltage On mode (01). It first turns on the ESA Low Voltage supply of each stack whose bit is set in the ESA power (ESA_PWR) entry in the ESA_ICT. Normally all bits are set, but a bit may be cleared before starting the IESALVON sequence to prevent a stack from being powered on. After a short wait (about 150 ms) for the power to stabilize, the ESA stack and formatter registers are set to their default values in the ESA_ICT, with the exception that all high voltage levels are set to zero and the high voltage enable bits are set off. At the end of the sequence, science telemetry is started.

3.5.2.2.2. Sequence 1: IESAHVON. IESAHVON can be run from either OFF mode or Low Voltage On mode, and transitions to the High Voltage On mode. If starting from OFF, the IESALVON sequence is first run. Next the high voltage sequence is run if the high voltage software enable key has been set (see Section 3.5.2.2.3 below). This sequence first sets all high voltage control registers in the ESA stacks to off. It sets the telemetry to 'diagnostic' mode (all ESA burst telemetry is channelled to survey memory, VC4, to ensure that it is not discarded), and turns on ESA Burst housekeeping to provide monitors of the high voltages for ground diagnostics while the supplies are turned on. The ESA High Voltage power switches are turned on, followed by a 150 ms delay. Next the sweep high voltage (SHV) supplies are enabled and ramped to their default values at a rate specified in the ESA_ICT. The ESA sweep supplies are next set to their default values. Then, the microchannel-plate (MCP) supplies are enabled and ramped up like the SHV supplies. Diagnostic telemetry mode ends, and the ESA burst housekeeping mode is changed to monitor the high voltage sweep waveforms for one spin, then turned off.

3.5.2.2.3. Sequence 2: IESAHVOFF. ESAHVOFF can be run from any state, but is nominally run from the High Voltage On state. It ends up in the Low Voltage On mode. The ability to run from any state provides a means of switching off the high voltages safely even if they were not turned on by the IESAHVON sequence (e.g., by direct CDI command). The sequence first sets the sweep voltages to zero, then the MCP supplies, then the SHV supplies. It then pauses for about 1 s for high voltages to discharge before setting all enables off and finally setting the ESA High Voltage power switches off.

3.5.2.2.4. Sequence 3: IESALVOFF. ESALVOFF can be run from any state, but is nominally run from the High Voltage On or Low Voltage On state. It ends up in the OFF state. The sequence first executes the IESAHVOFF sequence to be sure that the high voltages are off. Next all ESA telemetry is disabled. Finally, all ESA power switches are turned off.

3.5.2.2.5. Sequence 4: Pulse Height Distribution Test. This sequence measures the pulse height distribution generated by the ESA detectors, to determine the optimal MCP high voltage settings. It performs this by ramping the preamplifier Gain Digital to Analog Converter (GDAC) slowly while collecting data with the sweep voltages fixed at 110 V (sampling 900 eV particles). GDAC sets the threshold for MCP charge pulses that are counted. Thus the count rate reflects an integral of the pulse height distribution above a threshold set by GDAC. The test is repeated at five MCP high voltage settings centered at the current nominal setting. The five test cycles are run with MCPs set to nominal -100 V, nominal -50 V, nominal, nominal $+50$ V, and nominal $+100$ V. This test is intended to be done on-orbit with the high voltages on and the spacecraft in a stable plasma environment. High rate burst telemetry is collected and redirected to Survey memory (VC4) to ensure that the data is saved. On completion, the instrument is returned to the mode in which it started.

3.5.2.2.6. Sequence 5: Noise Test. This test checks the background noise levels to be sure that the preamplifier gain settings are correct. It is nominally designed to be run with high voltage off, so that only noise is measured, but the high voltage registers are not modified by this sequence. The instrument setup consists of redirecting Burst telemetry to Survey memory (VC4) to ensure that the data is saved. No other registers are modified. The GDAC is then ramped exactly as in the PHD test described above. At the end of the test, the high rate telemetry is returned to its nominal VC2 and the instrument is returned to its previous mode.

3.5.2.2.7. Sequence 6: Pulser Test. This test checks out the ESA electronics using the internal test pulsers. It also checks the GDAC threshold levels by ramping up the test pulser amplitude with the GDAC levels kept at their nominal levels. It is nominally designed to be run with high voltage off, so that only noise is measured, but the high voltage registers are not modified by this sequence. The instrument setup consists of redirecting Burst telemetry to Survey memory (VC4) to ensure that the data is saved. Next the test pulsers are enabled in fixed (non sweeping) full rate mode. The test Pulser gain Digital to Analog Converter (PDAC) is then ramped through a sequence of levels, with pauses for data collection between levels. The data collection delay is the same scheme used for Sequence 4 and 5.

At the end of the PDAC sweep, the test pulsers are changed from 'Fixed' to 'Sweep' mode, where the test pulse rate is proportional to the sweep high voltage (it does not matter if the high voltages have been enabled, although the sweep mode amplitude must be set above zero to get a reasonable pattern). It waits in this mode for data collection, then returns the registers to nominal values and stops. The sweep mode tests the relative timing between data collection and the instrument's energy sweep.

3.5.2.3. *High Voltage Enable.* The IESAHVON sequence is enabled by the software high voltage enable key. This key was implemented to prevent unwanted application of high voltage which could damage the electrostatic analyzers. This key can be set using the IESAHVEBL command with the proper KEY value as a parameter (command IESAHVEBL KEY= n). The key can be reset (to disable future attempts to turn on the high voltages) using the IESAHVDIS disable command. The default state for the High Voltage Enable key is disabled.

Note that even if the software enable key has been set, there are a number of other interlocks. A special CDI command to the power converter is required before power is sent to the ESA high voltages. In addition, the appropriate bits must be set in the power enable, high voltage enable, and high voltage DAC (Digital to Analog Converter) entries in the ESA_ICT. By default, the ESA_ICT registers are set for normal high voltage operation, so that all high voltages will come up in response to the IESAHVON sequence once the high voltage enable keys have been given to the software and the power converter.

3.5.2.4. *High Voltage Ramp Command.* This command can be used to ramp up the ESA high voltages once the instrument is in High Voltage On mode. This is the preferred way to change an MCP or SHV high voltage supply setting after the default has been set by the IESAHVON command. The command has three parameters. First the ESA stack name (ESA1, ESA2, ESA3, or ESA4); next the supply name (MCPA, MCPB, MCPC, MCPD, SHV1, or SHV2); and finally the desired setting (LEVEL= n). The level is set in DAC units (0–255). If the requested level is less that the current setting, the supply is reprogrammed to the new level in one step. If the requested level is greater than the current level, the supply is ramped up using the same scheme that is used for the IESAHVON transition sequence, which is controlled by entries in the ESA_ICT table.

3.5.2.5. *Diagnostic Telemetry Commands.* There are two diagnostic packet types (VC5) related to ESA operation: ESA burst housekeeping and CDI/ESA.ICT dump. They can be requested by command or can be generated periodically.

ESA analog housekeeping is normally reported with VC0 and VC1 housekeeping packets at a slow sample rate. Higher sample rates are sometimes desirable (e.g., during high voltage ramp-up, or to monitor the waveform on sweep supplies). The ESA Burst Housekeeping samples a selected ESA housekeeping parameter from each stack 1024 times per spin (about 200 samples per second). There are a number of different modes which determine how the ESA housekeeping channel select is controlled. The mode is reflected in the housekeeping packet header. ESA Burst Housekeeping is requested automatically during some transition sequences (e.g., IESAHVON) or is commandable with a selectable number of packets.

A diagnostic packet is generated periodically containing the current contents of all the IDPU control registers (CDI, 256 registers, 512 bytes), plus the ESA_ICT table (the first 512 bytes). The rate at which these packets come out can be set from once a second to every 65535 seconds (18 hours). This packet is primarily designed to diagnose problems, and to ensure that the state of the instruments is known for ground processing.

3.5.2.6. *Automatic Instrument Controls.* While in an operating mode, the ESA software performs a number of tasks to dynamically control the ESA analyzers, and to protect the instrument from faults. These tasks include sESA sweep voltage tracking, peak count rate and energy flux determination, ESA deflector control, and high voltage over-current protection.

3.5.2.6.1. sESA Sweep Voltage Tracking. The sESA (stepped electrostatic analyzer) tracking mode allows the 12 ESAs that generate the sESA data to perform high time resolution measurements of electrons over a limited energy range determined by a variable algorithm that uses the energy spectra from the eESA (electron electrostatic analyzer). The sESA Tracking mode involves having the Processor set the sESA sweep mode Top Energy step to track the energy at which the highest count rates is observed by the eESA. The Formatter contains a circuit to find the peak counts in the eESA data over a selected range of energies and anodes. The processor reads this data once per sweep, and passes the information to the sESA Tracking Module. This module then computes the new sESA Top Energy step, which can be set 'n' steps above or below the 'peak count' energy step, and reprograms the sESA sweep mode accordingly. In tracking mode, the sESA produces a 6 energy $\times$ 16 pitch angle measurement of a portion of the electron distribution function determined by the spectral peak. Both the tracking rate (number of eESA sweeps between sESA adjustments) and the energy spacing between the 6 pairs of sESA analyzers are selectable.

While in tracking mode, the sESAs can also be set to toggle between 2 energy steps producing a 12 energy step measurement with $\frac{1}{2}$ the time resolution, or to toggle between 4 energy steps producing a 24 energy step measurements with $\frac{1}{4}$ the time resolution. We note that the sESA was operated without tracking (fixed energy steps) during most of the first two years to facilitate analysis of long time periods, however most sESA operations during the following years have used this tracking.

3.5.2.6.2. Peak Detector. The formatter contains 'Peak Detector' circuits which find the energy at which the maximum count rate occurs in the eESA and iESA analyzers. Values in ESA_ICT are used to program the energy range and anode set that are used in the search. Anodes are grouped into 8 sets covering 45 deg each. The ability to select a piece of pitch angle space with the anode mask is desirable. This involves de-spinning the anodes, and rotating the anodes to align them with the magnetic field before selecting which anodes are to be used in the peak detector. The Formatter peak detector hardware does the anode de-spinning. (The de-spinning can be enabled or disabled, and the direction of de-spin controlled.) The de-spin is done to the closest anode at the start of the spin, and the rotation of the anode mask to align it with the magnetic field is done by software. Individual stacks can also be disabled from the peak detect computation. Peak Detector results are accumulated into a buffer for use by the Burst and Fast/Slow survey triggers. The values are also telemetered in the Triggers diagnostic packet.

3.5.2.6.3. Deflector Control. The eESA and iESA analyzers have deflectors designed to shift their Field of View (FOV) to include the magnetic field vector. Two deflectors are contained in each analyzer: one to deflect up and one to deflect down. The FOV can be deflected by up to about $\pm$ 7 deg (the spacecraft orientation should keep the magnetic field direction within these limits most of the time). Software controls the amplitude of the deflector setting, based on magnetic field information obtained from the science magnetometer. Hardware modulates this amplitude with the analyzer sweep voltage, since the deflector setting is proportional to the energy of the particles for a given deflection angle.

The deflector control computes the sign and amplitude of the deflector control setting relative to the sweep voltage, and programs the appropriate part of the eESA and iESA Sweep Mode registers. The deflector control module receives the magnetic field vector and magnitude (B1NT and BMAGNT) from the Fields Driver, and the result of the calculation is converted from 2's complement to sign-magnitude for the deflector control hardware and sent to the instruments.

3.5.2.6.4. High Voltage Over-current Trips and Other Protection. The software periodically checks the high voltage current monitors to ensure that the supplies are not exceeding a programmable current limit. If the current limit is exceeded for longer than a specified interval, then the supply is shut off. When a supply is shut off, a bit is set in the Fast Digital Monitor word. The time at which the most recent supply shut-down occurred is also recorded, and may be extracted from the ESA_ICT Diagnostic Telemetry packet. If a supply exceeds its limits for less than the specified time, a 'Glitch' bit is set for that supply which may also be monitored in the ESA_ICT Diagnostic Telemetry packet. The current trip limits are specified in the ESA_ICT. They default to about 20 mA for the MCP supplies, and 8 mA for the SHV supplies. The MCP supplies also have a special limit which applies when the MCP voltage setting is less than about 200 V. The trip current is much lower in this case, about 6 mA. This allows early detection of a shorted MCP supply, before the voltage goes very high.

In addition to current limit monitoring, there are some consistency checks made on the programmed high voltage settings. The high voltage optocoupler enables (see Carlson *et al.*, 2001, for high voltage hardware description) are switched off if the corresponding SHV supply is disabled. Also, if the SHV supply level is not high enough to support a programmed sweep top voltage step, the top voltage step is reduced. The ESA_ICT defines what minimum SHV setting is required to support a given sweep.

3.5.2.7. *Telemetry.* The ESA driver makes headers for all ESA science packets, diagnostic packets, and housekeeping packets. Science packet headers include a number of instrument configuration items. These are assembled synchronously with the start of data collection. The exception is the sESA Top energy, which is sampled several times during packet collection. The packet headers also include the particle mode setting from the ESA_ICT table. ESA Housekeeping packets include the status of the ESA high voltage enable registers and sweep modes (copied from the CDI status table), and the most recent readings of the ESA analog housekeeping (16 values per stack). The ESA analog housekeeping is sampled periodically, cycling through the samples roughly 8 times a spin (about 1.2 samples per second). The sweep voltages and currents are sampled synchronously with the spin so that the sample will always take place at the same part of the sweep.

3.5.2.8. *Routine Operations.* For routine operations, the commands required in Mode strings are typically a few commands to change some default values in the ESA_ICT, followed by the IESAHVON command. Typically, the Fast to Slow and Slow to Fast Survey commands change only the ESA Averager mode registers in the Formatter, by direct instrument CDI commands. The IESADIAG command may also be included to generate diagnostic packets more often during a pass. The mode command table should include a table load command to the ESA_ICT table to set the particles mode number, so that it may be reported in the telemetry packet headers.

3.5.3. *The TEAMS Driver*

The TEAMS mass spectrometer consists of an electrostatic analyzer combined with a time-of-flight velocity analyzer that together perform a 3-D measurement of the ion distribution function covering 4 ion masses (Klumpar *et al.*, 2001). Using spin-phase information generated by the IDPU, the TEAMS instrument sorts the incoming ion flux by spin phase, mass, and energy, and accumulates these events in a local memory. IDPU flight software transfers accumulated data from the TEAMS local memory through the formatter card and into the IDPU Mass Memory. In the process, the formatter card provides '16-bit to 8-bit', '14-bit to 8-bit' counter compression, and 'no' compression options depending on the type of data being transferred.

TEAMS table loads, time-of-flight high voltage setting and other instrument commands are serviced directly by CDI commands, while instrument analog levels are read back in 'slow' housekeeping (Section 3.2.2). This leaves the driver the functions of: (1) ramping up the MCP voltage at the start of each pass, (2) selecting which area of the TEAMS memory to transfer, (3) coordinating the transfer and (4) formatting the header for the resulting data.

The TEAMS instrument provides 8 data product types at a variety of rates controlled by internal registers. To keep the readout pattern in sync with the accumulation, the software driver uses the current spin phase plus identification in the particle mode commands to identify one of 13 required readout patterns and to determine which TEAMS memory page to read. The page number is then sent via a CDI command to the TEAMS unit before the transfer is initiated in the formatter card. Worst-case delays in CDI latency and the large VC2 data volume combine to make the TEAMS channel operate nominally at 95% of capacity, limiting the spin period to 4.7 s or greater.

The TEAMS interface presented an unexpected challenge as the only data generator which provided a mixture of VC2 and VC4 data. Since the hardware Mass Memory pointers in the formatter card are double-buffered, the TEAMS driver must predict what VC memory is going to be used in the next transfer before starting the current transfer. This is implemented by maintaining a simple 3-element transfer FIFO. In order to verify that all TEAMS data types in all 13 readout patterns were read in the proper phase despite FIFO-induced variations, driver logic had to be prototyped and characterized using a ground simulator.

4. Summary

The FAST satellite requires an intelligent, flexible Instrument Data Processing Unit (IDPU) in order to operate the large array of instruments and to collect the high rate data required by the scientific objectives. The IDPU controls and routes commands to the various instruments and electronics boards, controls the power system, oversees instrument housekeeping and boom deployments, directs data to the mass memory, generates data headers, and manages the data selection and triggering algorithms that optimize the data. A design philosophy was chosen to develop autonomous instruments that require minimal commanding and a data recording system that minimizes processor interaction. Custom designed field programmable gate arrays perform numerous tasks normally delegated to the processor. With a reduced work load for the processor, the software focuses on data optimization schemes that maximize the science return. The innovative design of the FAST IDPU has been essential to the success of the mission. It is hoped that this description can be used in the future to help integrate multiple spacecraft instruments into a single experiment.

References

Carlson, C. W., McFadden, J. P., Turin, P., Curtis, D. W., and Magoncelli, A.: 2001, 'The Electron and Ion Plasma Experiment for FAST', *Space Sci. Rev.* (this issue).

Ergun, R. E., Carlson, C. W., Mozer, F. S., Delory, G. T., Temerin, M., McFadden, J. P., Pankow, D., Abiad, R., Harvey, P., Wilkes, R., Primbsch, H., Elphic, R., Strangeway, R., Pfaff, R., and Cattell, C. A.: 2001, 'The FAST Satellite Electric Field and Magnetic Field Instrument', *Space Sci. Rev.* (this issue).

Klumpar, D. M., Möbius, E., Kistler, L. M., Popecki, M., Hertzberg, E., Crocker, K., Granoff, M., Tang, L., Carlson, C. W., McFadden, J., Klecker, B., Eberl, F., Kunneth, G., Kastle, H., Ertl, M., Peterson, W. K., Shelley, E. G., and Hovestadt, D.: 2001, 'The Time-of-Flight Energy, Angle, Mass Spectrograph (TEAMS) Experiment for FAST', *Space Sci. Rev.* (this issue).

McFadden, J. P., Ergun, R. E., Carlson, C. W., Herrick, W., Loran, J., Vernetti, J., Teitler, W., Bromund, K., and Quinn, T.: 2001, 'Science Operations and Data Handling for the FAST Satellite', *Space Sci. Rev.* (this issue).

Pankow, D., Besuner, R., Ullrich, R., and Wilkes, R.: 2001, 'Deployment Mechanisms on the FAST Satellite: Radial Wires, Stiff Axials, and Magnetometer Booms', *Space Sci. Rev.* (this issue).

Pfaff, R. C., Carlson, C., Watzin, J., Everett, D., Gruner, T., Curtis, D., Pankow, D., Heetderks, H., Harvey, P., Ergun, R., and McFadden, J.: 2001, 'An Overview of the Fast Auroral SnapshoT (FAST) Mission', *Space Sci. Rev.* (this issue).

MAGNETIC FIELD INSTRUMENTS FOR THE FAST AURORAL SNAPSHOT EXPLORER

R. C. ELPHIC
Los Alamos National Laboratory, Los Alamos, New Mexico, U.S.A.

J. D. MEANS, R. C. SNARE, R. J. STRANGEWAY and L. KEPKO
Institute of Geophysics and Planetary Physics/UCLA, Los Angeles, California, U.S.A.

R. E. ERGUN
Laboratory for Atmospheric and Space Sciences, University of Colorado, Boulder, Colorado, U.S.A.

Abstract. The FAST magnetic field investigation incorporates a tri-axial fluxgate magnetometer for DC and low-frequency (ULF) magnetic field measurements, and an orthogonal three-axis searchcoil system for measurement of structures and waves corresponding to ELF and VLF frequencies. One searchcoil sensor is sampled up to 2 MHz to capture the magnetic component of auroral kilometric radiation (AKR). Because of budget, weight, power and telemetry considerations, the fluxgate was given a single gain state, with a 16-bit dynamic range of ±65536 nT and 2 nT resolution. With a wide variety of FAST fields instrument telemetry modes, the fluxgate output effective bandwidth is between 0.2 and 25 Hz, depending on the mode. The searchcoil telemetry products include burst waveform capture with 4- and 16-kHz bandwidth, continuous 512-point FFTs of the ELF/VLF band (16 kHz Nyquist) provided by a digital signal processing chip, and swept frequency analysis with a 1-MHz bandwidth. The instruments are operating nominally. Early results have shown that downward auroral field-aligned currents, well-observed over many years on earlier missions, are often carried by accelerated electrons at altitudes above roughly 2000 km in the winter auroral zone. The estimates of current from derivatives of the field data agree with those based on flux from the electrons. Searchcoil observations help constrain the degree to which, for example, ion cyclotron emissions are electrostatic.

1. Introduction

The primary scientific thrust of the Fast Auroral Snapshot Explorer mission is to measure, characterize and understand rapidly changing and small scale microphysical processes associated with the aurora. Consequently the payload is designed to obtain very high-time resolution particle and fields measurements in low-altitude polar earth orbit. Measurements are made in the auroral zone between 350 and 4200 km altitude, and allow study of the plasma physics of auroral processes on length scales down to tens of meters. DC magnetic field measurements are essential for organizing particle distributions. They also provide an important large-scale backdrop for high sample rate and burst mode particles and fields observations. Magnetic field measurements in the auroral zone are critical for providing the instantaneous main earth's field for particle pitch-angles, for indicating the presence

Space Science Reviews **98:** 151–168, 2001.

of large- and small-scale field-aligned currents in the auroral zone and electromagnetic wave phenomena associated with auroral processes. The signatures of field-aligned currents are observed largely as east-west deflections in the background field, and are associated with north-south variations in the electric field at ionospheric altitudes (Sugiura *et al.*, 1982). This relationship was observed to hold down to the smallest scales sampled by the Dynamics Explorer spacecraft at low altitudes (Weimer *et al.*, 1985). The FAST magnetometers thus have to provide information in both the large-scale/low frequency regime and the small-scale/high frequency regime.

2. Measurement Requirements

Factors influencing the dynamic range of magnetometer design are the need to measure the earth's full field as well as auroral structure down to arc scale-lengths. Thus field variations range from $\pm 65\,000$ nT through the largest scale currents ($\sim$1000 km) which typically cause deviations of several hundred nT at a few hundred kilometers altitude. At the smallest scales, Dynamics Explorer measurements indicate current-related structure down to a few nT. These extremes were discussed by Weimer *et al.* (1985), who showed electric and magnetic fluctuation amplitude as a function of equivalent wavelength down to about 5 km. The amplitude at this shortest wavelength is 3–5 nT; it is desirable to measure down to $\pm$ 2 nT to capture the smaller scale structure.

To resolve the smallest scale auroral current structures (100 m and less; cf., Borovsky (1995)), the field must be sampled at rates consistent with the convection of these structures past the spacecraft. Assuming a fastest-case convection speed (satellite plus arc motion, normal incidence) of 10 km s^{-1}, a structure of 100 m wavelength corresponds to a Nyquist frequency of 100 Hz, or a sample frequency of 200 Hz. Auroral zone ELF observations from Freja indicate background spectral densities of about 10^{-6} nT^2 Hz^{-1} at 80 Hz (between the O^+ and He^+ gyrofrequencies at Freja's altitude), and about 10^{-8} nT^2 Hz^{-1} near 1000 Hz (Erlandson *et al.*, 1994). Gurnett *et al.* (1984) have shown that low-altitude auroral zone magnetic spectral densities at 100 Hz are roughly 10^{-7} nT^2 Hz^{-1}, with a spectral slope much steeper than f^{-1}. At 100 Hz, background fluctuation amplitudes are about 1 pT. To study structures of 10-m scales, spectral densities below 10^{-10} nT^2 Hz^{-1} at 1 kHz must be measured.

Additional sensitivity requirements arise from the need to measure ion cyclotron waves, chorus and auroral kilometric radiation. The last is a stressing measurement objective, aimed at acquiring electromagnetic wave data at the lower end (several hundred kiloHertz) of the auroral kilometric radiation band.

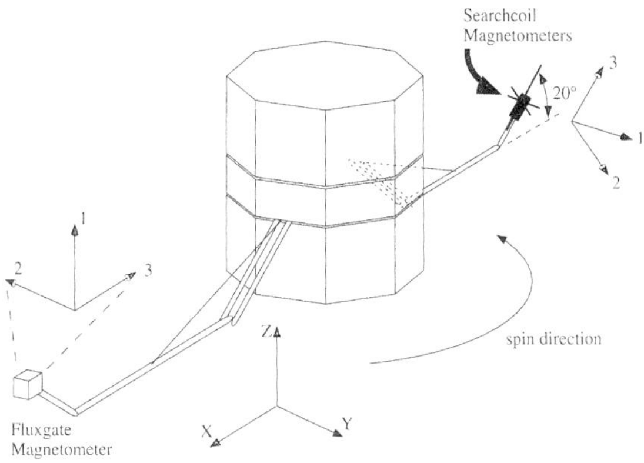

Figure 1. Schematic of the FAST DC fluxgate and AC searchcoil magnetometer mechanical configuration, as deployed in flight. See Pankow *et al.*, this issue, for details on the boom design.

3. Magnetometer Design

The necessary wide dynamic ranges in sensitivity and sampling frequency for FAST led us to use two techniques in magnetic field measurement. The background field, and large- and medium-scale field-aligned current signatures is most suited to a fluxgate magnetometer. For reasons made clear below, a search coil magnetometer is best for studying the higher frequency electromagnetic waves. This is primarily due to the sensitivity of the two designs at different frequency regimes; we intended that the two magnetometers overlap in frequency coverage.

The FAST triaxial fluxgate magnetometer heritage includes the OGO 5, ISEE 1 and 2, Pioneer Venus Orbiter, AMPTE UKS, and Galileo magnetometers, but it draws nearly all of its analog design from the Polar Magnetic Field Experiment (Russell *et al.*, 1995), so that development costs were kept to a minimum. Earth's large background field demands a range of ±65 000 nT. With the use of a 16-bit, low mass, low power flight qualified analog-to-digital converter (ADC), we obtain a resolution of ± 2 nT, sufficient to resolve auroral current structures at large (100's of km) to small (1 km) scale. For a Nyquist frequency of 100 Hz, digitization noise is $(2\ \mathrm{nT})^2/12F_N = 3.3 \times 10^{-3}\ \mathrm{nT}^2\ \mathrm{Hz}^{-1}$. As discussed above, the level of natural signals can be several orders of magnitude lower. Consequently the useful fluxgate bandwidth at 16-bits resolution (for ±65 000 nT) is from DC to about 20 Hz at FAST altitudes.

In order to measure electromagnetic waves and structures approaching the few tens of meters scale size, we use a triaxial search coil design. All three sensors cover the range from about 10 Hz to 16 kHz, and one of the three has a higher frequency response, out to 2 MHz to capture the low end of AKR. This approach provides overlap between the fluxgate and search coils at low frequencies. The triaxial design allows us to completely specify wave polarization and, with the electric field instrument, the Poynting flux of electromagnetic waves.

The deployed fluxgate and searchcoil mechanical configurations on FAST are shown in Figure 1. The three fluxgate sensors are sampled and digitized in the order shown; sensor 1 nominally points along the spin axis when the boom is properly deployed (see Pankow *et al.*, this volume), sensor 2 points opposite to the motion of the boom about the spin axis, and sensor 3 points back toward the spacecraft, along the boom. The fluxgate boom azimuth defines the X-axis of the spacecraft coordinate system.

The fluxgate magnetometer is a three axis instrument using highly stable low noise ring core sensors. The sensors are boom mounted approximately two meters from the spacecraft body, in a shielded housing $114 \times 54 \times 61$ mm in dimension. The sensors, housing, mounting bracket hardware and pigtail weigh 420 gm. The triaxial search coil magnetometer consists of two 7-inch (178 mm) sensors, and one 21-inch (533 mm) sensor mounted orthogonally. The 7-inch sensors have cores of laminated molybdenum permalloy, wrapped with a coil of 10 000 windings on a fiberglass bobbin. The 21-inch sensor consists of a laminated amorphous metal core laid in a fiberglass support structure, wrapped with a coil of 2000 windings on a fiberglass bobbin. One 7-inch sensor (sensor 1) is mounted so as to lie in the spin plane, the other (sensor 2) roughly antiparallel to the spin axis but canted outward from the spacecraft by 20°, and the 21-inch sensor (sensor 3) points approximately along the boom, but 20° up out of the spin plane. This mechanical configuration (see Figure 1) results from several constraints: (1) Mechanical stiffness and support during launch vibration is enhanced. (2) The 21-inch sensor fits within the Pegasus payload fairing. (3) The sensors have a full 2-m distance from the spacecraft following boom deployment. (4) The off-spin axis canting provides a measurement every spin of the vector AC magnetic field for higher frequencies, including AKR. The sensor assembly measures $533 \times 178 \times 178$ mm, with shielded pre-amplifiers located at the sensors, and weighs 707 g.

A single 251×257 mm circuit board contains the electronics needed to provide instrument clock, drive, sense and feedback to the fluxgate sensors, as well as Nyquist filtering of the analog voltages that are handed off to the switching and ADCs of the fields processor. This board also handles second stage amplification and filtering of the search coil signals, and hands them off to the fields switching/ADCs. It weighs 788 g. The entire analog magnetometer package consumes approximately 1.5 W.

Figure 2 is a block diagram of the analog magnetometer package. The fluxgate sensor electronics provide a 9 kHz drive signal for the sensors, a feedback signal which keeps the ring cores in zero field, and detect and amplify the second harmonic signals that are proportional to the ambient magnetic field. There is a 200 Hz low pass filter on the output, and the filtered result is provided to a switching system instead of a multiplexer (cf., Ergun *et al.*, this volume). This design provides greater isolation and less cross-talk between fluxgate channels; voltages settle better before being converted. The switching system output is provided to one of the fields experiment 16-bit ADCs (Ergun *et al.*, this volume). To simplify design

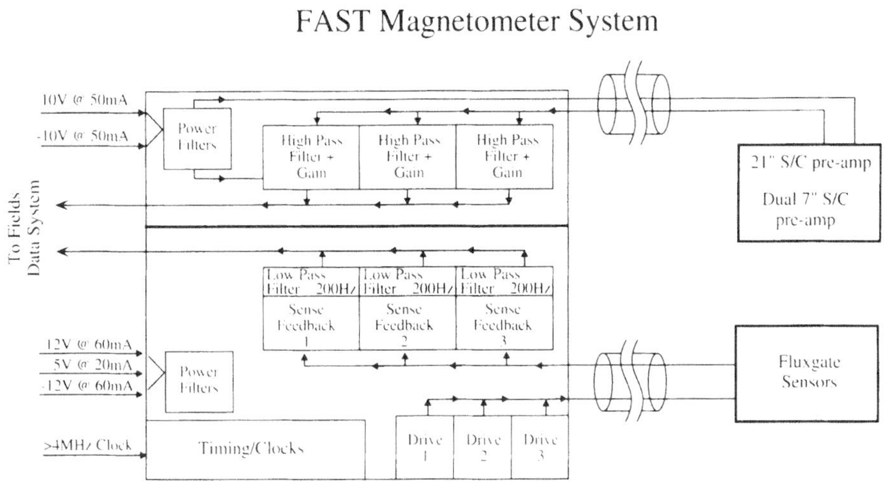

Figure 2. Block diagram of the analog DC and AC magnetometer package.

and reduce costs, the fields ADCs are generic, so the fluxgate data are sampled in the same way as electric field data. All searchcoil sensors are high-pass filtered to reduce the low-frequency AC signal due to the spinning main magnetic field. Data handling from the fields experiment ADCs are described in detail elsewhere (Ergun *et al.*, this volume).

4. Calibration

Initial ground calibration of the fluxgate magnetometer was complicated by noisy coils in the Goddard Space Flight Center magnetic calibration facility. Consequently, preliminary calibration was done on-orbit against the very well-characterized terrestrial field. This is considered adequate for FAST, since the thrust of the mission is to study field variations, not map the geomagnetic field to high accuracy and precision. The preliminary calibration resulted in a coupling matrix, which transforms data from sensor units in the non-orthogonal sensor system to geophysical nT quantities in the spinning spacecraft system. The preliminary coupling matrix is:

$$\begin{pmatrix} 0.984672 & 0.000000 & -0.008604 \\ 0.008272 & 0.928669 & -0.014153 \\ 0.006700 & 0.006234 & 0.957579 \end{pmatrix}.$$

The searchcoil sensors were calibrated on the ground in an end-to-end test using AC field stimulation by a solenoidal coil within a mu-metal flux can. The amplitude and phase relationship of the three sensors to the input signal was recorded at different frequencies. The two 7-inch sensors have peak measured sensitivities of 1.9 V nT^{-1} Hz^{-1} at 8800 Hz, while the 21-inch sensor has a higher frequency response and greater peak sensitivity, 26.8 V nT^{-1} Hz^{-1} at 60 kHz. The detailed transfer functions of the search coil and fluxgate sensors are discussed below.

5. Command and Data Handling

Both the fluxgate and search coil instruments have single gain states; no ranging is required. Commanding involves providing power to the instruments, and in selecting the rate and type of data acquired in the fields package. The fastest fluxgate analog front-end sampling runs at a rate of 2048 samples s^{-1}, but at lower telemetry volumes, spot samples are taken at rates that are powers of 2 smaller, down to 16 samples s^{-1}. Following conversion, a digital recursive anti-aliasing filter is applied to the data, and the results are decimated by a factor of four (meaning that one in every four samples is kept). The output rate of the fields processor for DC fluxgate data are varied in different modes (see Ergun *et al.*, this volume) as required by survey and burst telemetry volume needs. The recursive filter ensures signal fidelity at the lowest telemetry rates during slow survey and non-auroral zone background data modes (output rates typically 4 or 8 samples s^{-1}). The fluxgate data are telemetered only as 16-bit waveforms, with output sample rates from as much as 512 samples s^{-1} down to 4 samples s^{-1}.

The fluxgate magnetometer sensors are sampled nearly, but not quite, simultaneously. The fields switching system samples sensor 1, then 2, then 3, then pauses a beat before starting again. The inter-sampling time is long enough to allow the voltage to settle, and is 5/16 of the overall sensor 1-2-3 sample cycle time. This basic sample rate, as described above, ranges from 2048 cycle s^{-1} down to 16 cycle s^{-1}, depending on fields data acquisition mode.

The digital recursive filter takes 1/16 of the current sampled magnetometer output from the ADC, and adds it to 15/16 of the previous filtered output value, and stores the result as the current output of the filter. These stored values have 20 bits of precision. When decimated by four and sent to the packetizer, they are truncated to 16 bits. The effective transfer function of the FAST fields processor data acquisition for the fluxgate magnetometer is

$$T(f) = (\tfrac{1}{15})/(\tfrac{16}{15} - e^{-i\pi f/4F_{No}}), \tag{1}$$

where f is frequency and F_N is the Nyquist frequency of the original (undecimated) samples, four times higher than the effective output Nyquist frequency. This is shown in Figure 3; the 3 dB point is at about 0.15 F_{No}, the output (decimated) Nyquist frequency. The fluxgate absolute frequency response depends on telemetry mode, and is summarized in Table 1; the recursive filter corner frequency (approximately the effective bandwidth) can be as low as ~0.2 Hz, and goes as high as ~22 Hz in some auroral fast survey modes.

TABLE I

FAST DC magnetometer bandwidth

Output sample Rate (sps)	Output Nyquist frequency F_{No} (Hz)	Filter Corner frequency (Hz)
512	256	21.8
256	128	10.9
128	64	5.45
64	32	2.72
32	16	1.36
16	8	0.68
8	4	0.34
4	2	0.17

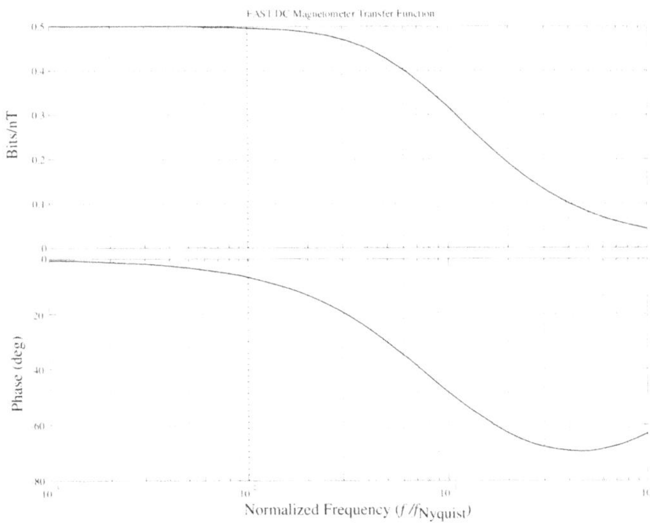

Figure 3. Effective transfer function of the DC magnetometer system. The transfer function is fixed to the output Nyquist frequency, which for FAST ranges from 2 to 256 Hz.

The search coil data are handled in several ways: waveform data are provided in survey mode at rates of from 32 to 2048 sample s^{-1} depending on telemetry mode, and in burst modes with 4-kHz bandwidth from all three sensors, and with 16-kHz bandwidth from the 21-inch sensor; high speed burst data result in waveforms with 1 MHz bandwidth from the 21-inch sensor; 512-point FFTs (from 0 to 16 kHz) of the 21-inch sensor data are performed by a DSP chip on orbit and passed down at various rates in specific packets; finally, swept frequency analyzer (SFA) data is captured with a bandwidth of 1 MHz. The searchcoil magnetometers are high-pass filtered to attenuate the Earth's main field at the spin frequency. The corner frequencies of the filters are 1290 Hz for the 7-inch sensors and 201 Hz for the 21 inch sensor. The transfer function of the searchcoil magnetometers is shown in Figure 4a for the 4-kHz bandwidth burst quantities, and in Figure 4b for the 16-kHz bandwidth burst quantity from the 21-inch searchcoil. The 180° phase relationship at low frequencies results from a combination of the high-pass filter, which produces a 90° phase shift at frequencies well below the corner frequency, and the basic searchcoil response to $\partial B/\partial t$, another phase shift of 90° from B. Likewise, the sensitivity (bits nT^{-1}) increases as f^2 for frequencies below the high-pass corner frequency. Above this frequency, the phase moves toward the expected value of 90°, and sensitivity (bits nT^{-1}) goes up as f.

6. Data Reduction

Data reduction approaches depend on handling either waveform or spectral data. As described above, the output fluxgate sensor samples must be properly time-aligned, then deconvolved from the filter response, then orthogonalized and calibrated to geophysical units. The signal is reconstructed by interpolating the three missing values and then inverting the filter. This process necessarily amplifies any noise at higher frequencies, so we typically cut off the filter inversion at 10% of F_{No}, depending on the strength of the natural signal.

The searchcoil waveform data are samples taken at either 8192 or 32768 samples s^{-1}, depending on the burst waveform capture channel. The voltages digitized by the ADC are proportional to nT-Hz, and so must be calibrated taking into account the high-pass analog filter as well as the intrinsic linear gain in frequency. Pre-launch tests provided detailed transfer functions for the three sensors, and their response can also be checked on orbit by examining the amplitude and phase of the signal caused by the main field at the spin frequency. Waveform data are integrated first to recover from the high-pass filter, then again to transform from a voltage proportional to nT-Hz to nT.

Spectral data from the searchcoils undergo the same calibration on the ground, but this time in the frequency domain.

Attitude determination is critically important to reducing the magnetometer data. Spin axis pointing is controlled by a magnetotorquer which is activated as necessary near perigee, where magnetic fields are largest and the torque authority is greatest. Spin rate is maintained by a second magnetotorquer. Knowledge of spin axis pointing is determined roughly by means of a spinning sun sensor, which measures sun elevation angle and sun-crossing time. A much more refined spin axis pointing is obtained by using the measured magnetic field from a full orbit and comparing this to the IGRF 1995 model. Using a full orbit's worth of spin axis sensor measurements helps constrain the spin axis to a small fraction of a degree. The spinning sun sensor also provides absolute spin phase via a sun pulse when the

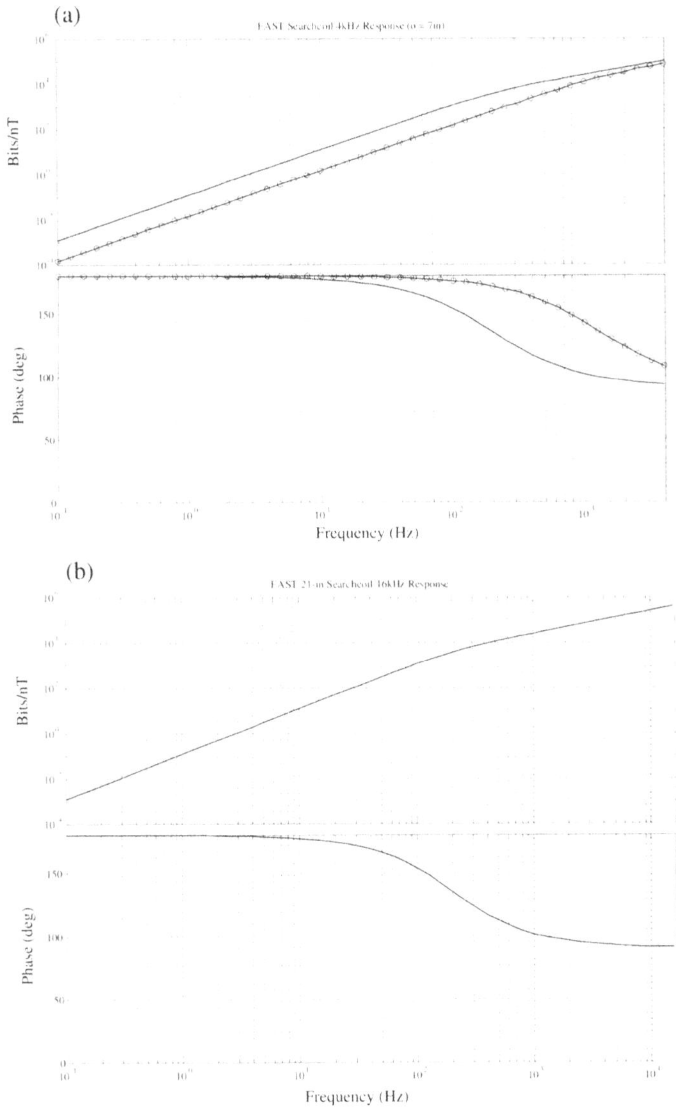

Figure 4. (a) Transfer function of searchcoils for the 4096-Hz bandwidth burst quantities. (b) Transfer function of the 21-inch searchcoil (Mag3ac) for the 16384-Hz bandwidth burst quantity.

satellite is in sunlight, while infrared horizon-crossing sensors provide coarser spin phase in eclipse. However, because of jitter in these pulse times the instantaneous absolute phase is poorly constrained. Consequently we smooth the roll phase but require it to pass through zero phase within a 'jitter window' of the nominal pulse time. This approach results in a smooth despinning process.

Relative gains and offsets for the spin plane sensors are obtained through a straightforward analysis of the DC and AC signals at the spin frequency performed

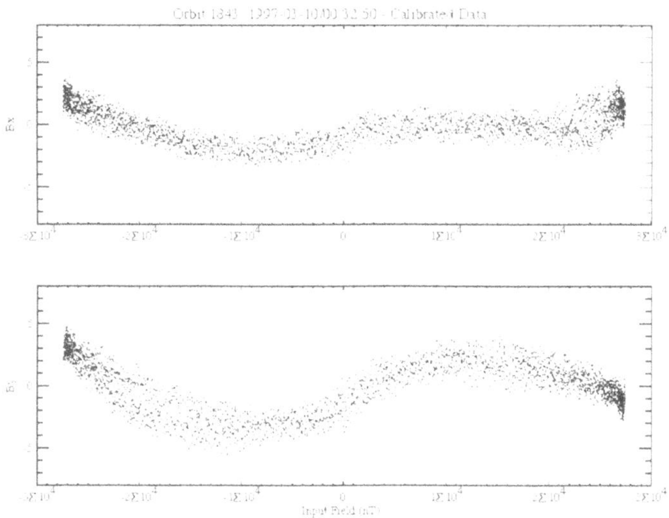

Figure 5. FAST fluxgate magnetometer nonlinearities. Shown here are the residuals from a sine wave fit to the X and Y axis data over several spins. The residuals show the sensor's departures from an ideal sine wave.

on every 2-spins worth (10 s) of data. The model used for each sensor is $(a_0 + a_1 t)\cos\omega t + (b_0 + b_1 t)\sin\omega t + c_0$, which allows for a linearly growing or decreasing background field amplitude over the 10 s interval, plus a constant value (offset + spacecraft field). Correlation of the spin plane sensor c_0 values with the spin axis field provides a measure of how much the spin plane sensors are canted out of the spin plane. A similar analysis of the spin axis sensor provides a measure of the off-spin cant angle. When this analysis is combined with the constraints of the IGRF model field, we can determine the parameters necessary to orthogonalize the sensors and determine absolute gains and offsets as well. Thus, adjustments to the relative calibration can be made smoothly every ~10 s as the data are processed. A correction for absolute calibration is obtained by comparing a full orbit of magnetometer data to the model field on *Dst* quiet days, about once a week.

7. Instrument Operation on Orbit

The fluxgate and searchcoil magnetometers have functioned normally for much longer than the nominal 1-year mission duration since turn-on in August of 1996. The boom deployment shortly after launch was flawless and placed the sensors in their nominal flight mechanical configuration.

The fluxgate magnetometer exhibits small departures from a linear response when spinning in Earth's main field at FAST's spin period of 5 s. The nonlinearities are first noticeable around 10 000 nT. Figure 5 shows the residuals from a sine wave fit to the sensor data over two spins of the satellite, plotted as a function of the input field as modeled by the fit. The top panel shows the residuals for the X-sensor, sensor 3, the lower panel for the Y-sensor, sensor 2; a perfectly linear response would result in a flat line. The sensor 3 (MagX) response is linear within 2 nT out to about 10 000 nT, the overall departures from linear are about 3 nT in 30 000, or 10^{-4}. The sensor 2 response is somewhat worse, about 2×10^{-4}. We were also able to determine the spin axis sensor behavior before boom deployment (when the booms are stowed, sensor 3 is antiparallel to the spin axis, and sensor 1 lies parallel to the spin plane), and its nonlinear character was smaller than 10^{-4}. Since sensor 1 lies nearly along the spin axis after deployment, its spin frequency AC signal is much smaller than 10 000 nT, and its response is effectively linear. For the deployed spin plane sensors, the nonlinearities are not a problem because (a) they are small and (b) they are so systematic they can be easily removed with a simple function.

The fluxgate magnetometer boom undergoes very small, but detectable, changes in orientation over the course of an orbit. In Figure 6, we plot an orbit's worth of four quantities: the cosine and sine of the angle between the two spin plane sensors (TW_YX, TW_YY) multiplied by the gain ratio of the two sensors (ideally unity), and the angles between the spin axis sensor and the X spin plane direction (TW_ZX) and the Y spin plane direction (TW_ZY). The first two quantities describe how the spin plane sensor mechanical mounting and/or electronic cross-talk change over the course of an orbit; the second two quantities provide similar information on mounting/cross-talk for the spin axis sensor, but also provide some measure of boom flexure and torsion. The satellite leaves eclipse at about 07:27 UT, moves across Earth's dayside toward perigee at about 09:00 UT, and enters eclipse again around 09:04 UT. At eclipse exit, there is little or no change in the spin plane sensors alignment/cross-talk, but there is a rapid change in TW_ZX from 0.3 milliradians to 0. At the same time, TW_YX exhibits no rapid change but begins a gradual trend from 0 to 0.5 mradians over about 20 min. These diagnostics indicate that there is a very prompt response ($\tau < 2$ min) by the magnetometer boom to the sunlight/eclipse transition, and a longer time-scale ($\tau \sim 20$ min) for torsion or twisting of the boom. Since these diagnostics are obtained by detailed analysis of the spin frequency signals on all three axes, it might be argued that some of the apparent spin axis sensor changes at eclipse entry and exit are due to a signal caused by solar array currents flowing while the satellite is in sunlight. Such currents would produce, to lowest order, a spinning dipole field as seen by the magnetometer sensors. However, such a contribution would be roughly constant in amplitude, while the signal we observe is instead a constant fraction of the magnitude of the spin plane field, and hence is an angular displacement. It can be seen that a change of similar size in TW_ZX but of opposite sign takes place as the satellite moves from sunlight to shadow at around 09:04 UT. A change in the apparent zero level of the X spin plane sensor (which should be proportional to the

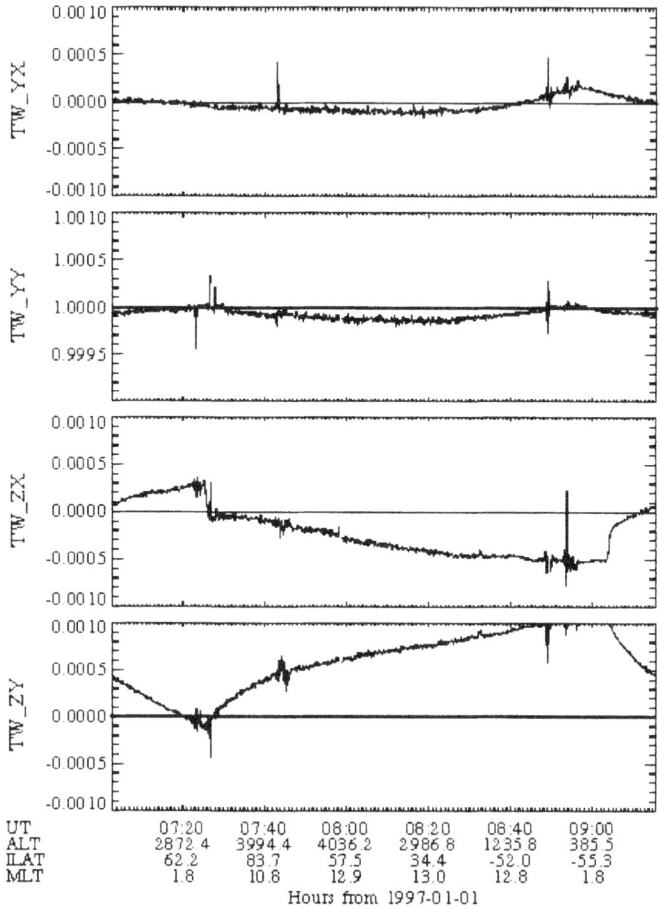

Figure 6. FAST fluxgate magnetometer orthogonality diagnostics, showing small changes in sensor and/or boom orientation with respect to the spin axis and spin plane over the course of an orbit. Exit from eclipse is at 07:27 UT, entry is at 09:04 UT, a few minutes after perigee. The diagnostics indicate that boom flexure and torsion are small but detectable, in the 0.3–0.5 mradian range (0.02° –0.04°).

change in the sensor's out-of-spin-plane angle) confirms that this is a boom flexure effect.

The FAST spacecraft is magnetically very clean. Except when the torque coils are operating, spacecraft fields are negligible. In any case we cannot distinguish between spacecraft fields and the intrinsic magnetometer offsets, which are fairly small and stable. The sum of offsets and spacecraft fields vary slowly with time, and exhibit a mild temperature dependence. The torque coils, which operate infrequently, create fields at the fluxgate of about 500 nT. The only AC interference observed by the searchcoils is the fluxgate drive frequency at 9 kHz, and this signal is at a low level.

8. Science Results

One issue which is central to auroral physics specifically and magnetosphere/ionosphere coupling in general is how and where field-aligned currents close through the ionosphere and in the magnetosphere. FAST is uniquely suited to studying auroral microprocesses, but the particles and fields instruments also address questions of a more global nature, like the closure of field-aligned currents (FACs). Of course, FACs have been observed and studied by rockets and satellites for decades - the new observations which FAST brings to the discussion are high sensitivity, high time resolution plasma measurements with continuous monitoring of all pitch angles, and high time resolution measurements of fields.

The FAST fluxgate magnetometer observes small-scale regions of downward-going current within the large-scale field-aligned current regions of the auroral zone. Associated with these downward currents are upward flowing low- and medium-energy (up to about 1 keV) electron beams. The scale size of the downward currents can be kilometers or less, often concentrated at the edges of inverted-V structures. These upgoing electrons have occasionally been observed before (Boehm *et al.*, 1995; Clemmons *et al.*, 1995), but the FAST instrumentation allows continuous high-time-resolution monitoring of particle distribution functions. The upgoing accelerated electrons/downward currents are also associated with diverging electric fields, and presumably with the 'black aurora', the optical counterpart of bright discrete arcs (Marklund *et al.*, 1994, 1997a, b). FAST data taken at apogee (4000 km) and near magnetic midnight commonly show downward FACs with current densities of between 2 and 4 microamps per square meter, consistent with the observed upward flux of $\sim 10^9$ electrons cm^{-2} s^{-1}.

Figure 7 shows DC magnetic field and electron plasma analyzer data. The field data have been detrended using the model field, IGRF 1995, but are in despun spacecraft coordinates, with the top panel corresponding to most of the main field. The second panel shows the north-south excursions, and the third the east-west excursions – all to the same scale. The plots go from lower latitudes to higher latitudes, starting at about 67° ILAT and ending at about 73°. A positive excursion of ΔB_{EAST} corresponds to a downward current, a negative excursion to an upward current.

The bottom two panels show the electron energy spectrum between 5 eV and about 30 keV, summed over all pitch-angles, and the electron pitch-angle spectrum summed over all energies. 0° and 360° correspond to field-aligned, or downgoing electrons, while 180° is upgoing from the ionosphere. One can see the atmospheric loss cone between roughly 150 and 210°.

At lower latitudes, roughly isotropic plasma sheet electrons are observed, which at higher, sub-auroral latitudes are interspersed with upgoing accelerated electron beams. Each beam interval is associated with an increase in ΔB_{EAST}, corresponding to a downward FAC. Then FAST passes into the auroral acceleration region proper, with inverted-V structure, lasting about a minute during which the space-

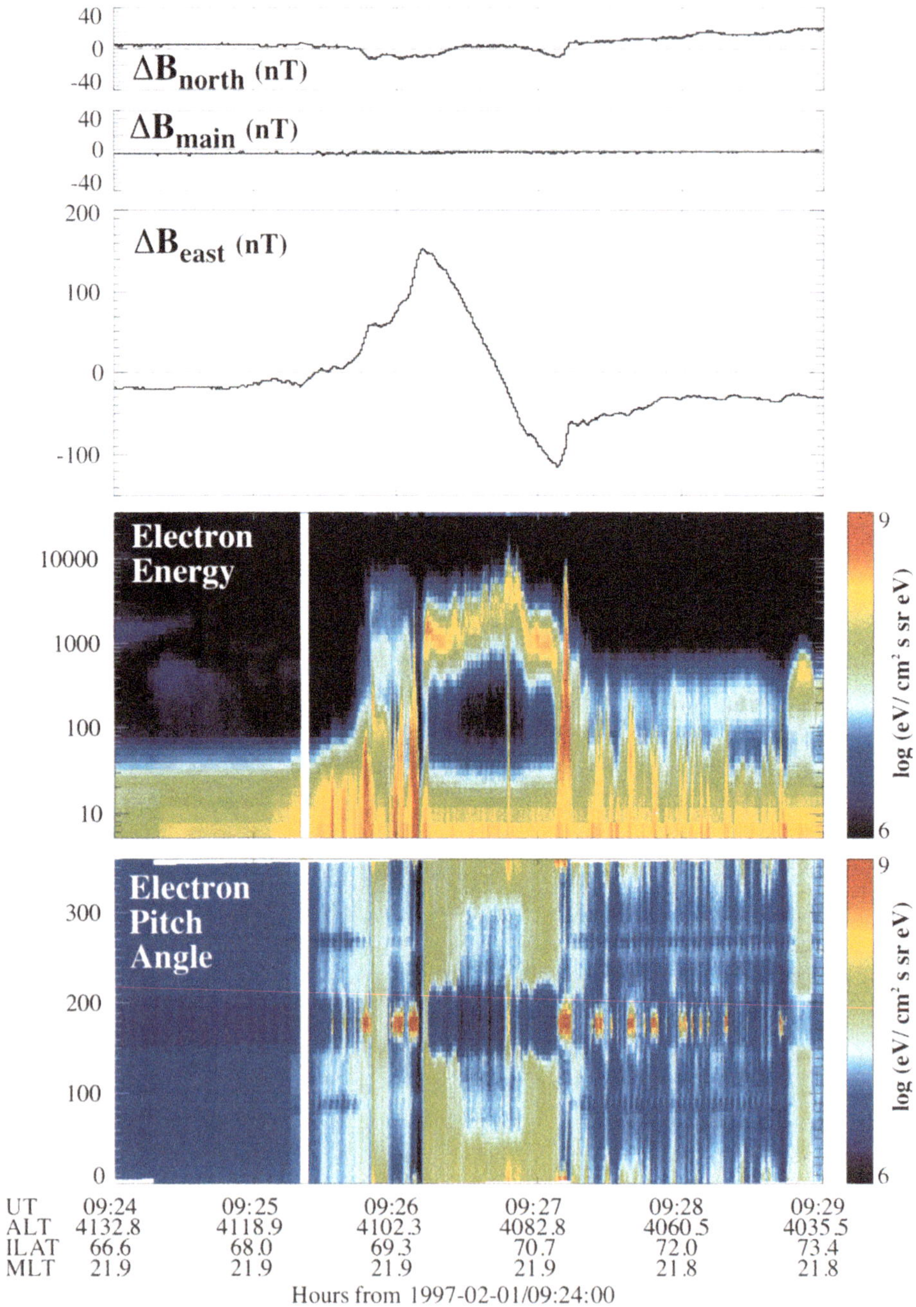

Figure 7. FAST fluxgate magnetometer and electron electrostatic analyzer data from a pre-midnight auroral zone pass on February 1, 1997. The magnetometer data have been detrended with IGRF 1995, but are in spacecraft coordinates – the top panel is essentially directed to the north, the second and third panels correspond to the main-field and east-west directions, respectively. The fourth and fifth panels are electron energy (over all pitch angles) spectra and electron pitch-angle (over all energies) spectra. Upward electron beams can be seen at pitch angles of 180°, and these correspond to positive deflections in the ΔB_{EAST} component, consistent with passage through downward field-aligned currents.

craft moves roughly 360 km. The field perturbation in this region is all upward FAC, as one would expect. This is followed by a brief but very intense upgoing electron beam, with energies going up to nearly 10 keV. The total downward current found equatorward and poleward of the inverted-V region just about balances the upward current carried by auroral precipitation, and by the time FAST has gone up to 73° the perturbation field has returned to nearly its starting point, indicating that little net current was flowing into or out of the auroral ionosphere on this occasion at this local time. Conditions for this interval were quiet, and the interplanetary field based on WIND key parameters was northward. Figure 8 displays some of the fields wave data products created onboard FAST (see Ergun *et al.*, this volume). Here we focus in on the AC electric and magnetic field observations within the primary auroral acceleration region for another FAST premidnight northern winter pass. Once again FAST is flying from lower latitudes to high, this time close to 23 MLT. The top panel shows the eastward magnetic field component, while the second and third panels provide the electron energy and pitch angle spectrograms that place the currents and waves in context. The next four panels are, respectively, high- and low-frequency spectrograms of the electric and magnetic field wave activity. High frequency in this case refers to the swept frequency analyzer data between roughly 10 kHz and 2 MHz, and the low frequency denotes the 16 Hz to 16 kHz DSP fourier analysis. The SFA spectrograms show the presence of auroral kilometric radiation near roughly 300 kHz, while the DSP spectrograms display a combination of VLF saucers, VLF hiss, ion cyclotron waves and lower hybrid emissions.

At about 20:50:10 UT, FAST was observing precipitating electrons with energies near 10 keV, carrying an intense upward FAC (note the steep negative slope in B_{east}), and collimated beams of upward-going ions (not shown) with energies of several keV. This situation obtains when the parallel accelerating potential is partly above and partly below the satellite. Within this interval, emissions at the hydrogen cyclotron frequency near 200 Hz can be seen in the electric and magnetic field DSP spectrograms. Figure 9 focuses in on a short period of time, 20:50:09.927–20:50:10.176 UT (denoted by the black vertical line in Figure 8), within one of the inverted-V's of the primary auroral acceleration region. Shown here are the 4096-Hz bandwidth waveforms from the three searchcoil sensors (see Figure 1) and the electric field (cf., Ergun *et al.*, this volume). The clearest signal is the ~200-Hz H^+ cyclotron frequency emission. Searchcoil sensors 1 and 2 were oriented at about 90° to the main magnetic field at this time, while sensor 3 was oriented roughly within 20° of the field. Consequently the difference in amplitudes at the three sensors is likely related to the polarization of the magnetic part of the wave. There are also lower-frequency fluctuations in sensors 1 and 2 that roughly correspond to the gyrofrequency of O^+ at ~12 Hz; but with amplitudes of ~1 nT, these waves would be at about the 7-inch searchcoil digital noise level near 10 Hz (see Figure 4(a)). Moreover, there is little counterpart signal at ~12 Hz in either the more sensitive 21-inch searchcoil (sensor 3) or the electric field waveform.

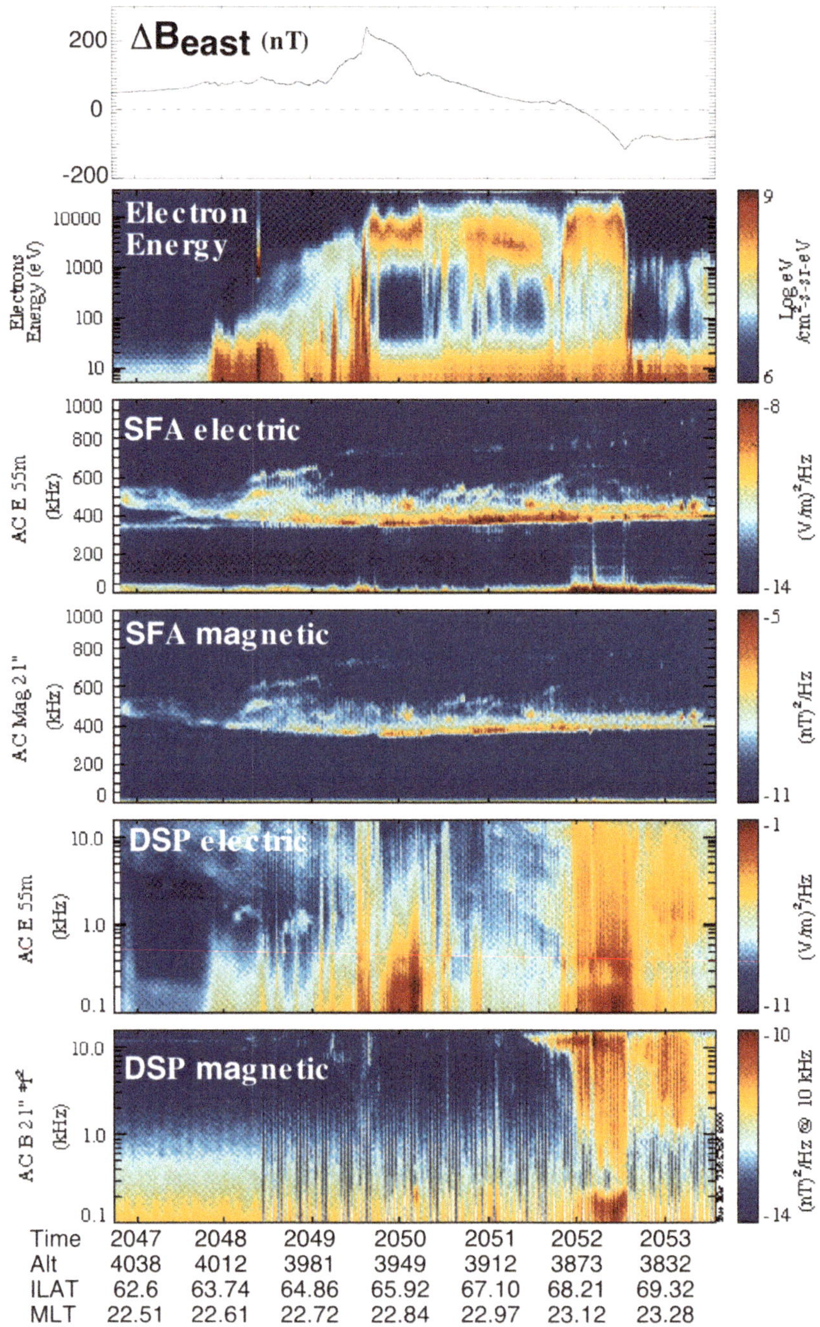

Figure 8. FAST ΔB_{EAST} magnetic field component, electron energy spectrum, and electric and magnetic SFA and DSP wave spectrograms for a pass on February 7, 1997. The electric and magnetic components of auroral kilometric radiation can be seen near 300 kHz in the SFA spectrograms. The DSP spectrograms display VLF saucers, hiss and ion cyclotron emissions.

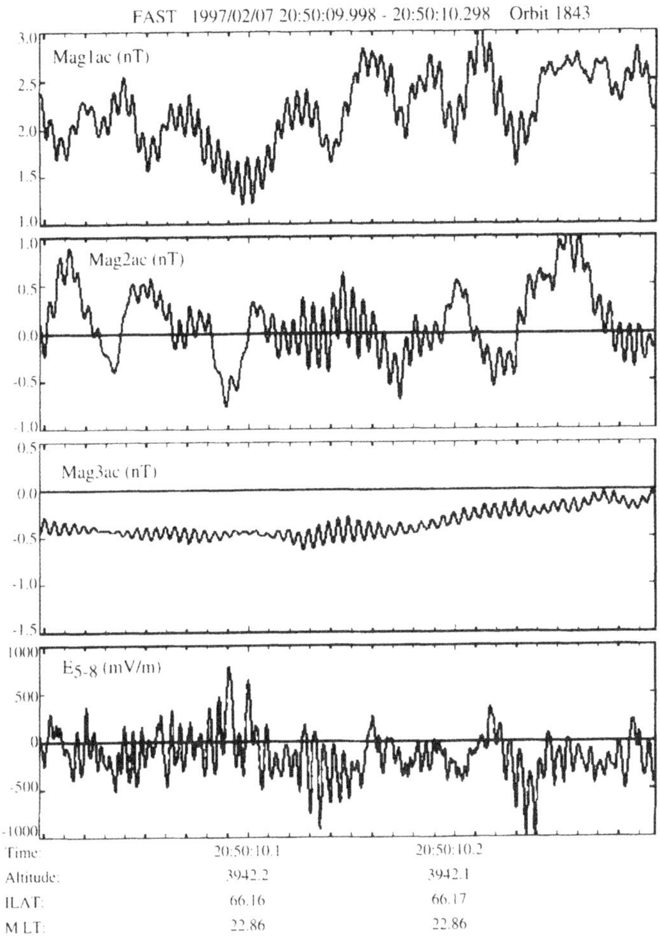

Figure 9. FAST 4096-Hz bandwidth waveform data for the three searchcoil sensors and the electric field for a period of about 250 ms from Figure 8. The principal feature is a hydrogen cyclotron emission at ~200 Hz.

9. Summary

The FAST magnetometer investigation consists of a tri-axial fluxgate magnetometer to measure DC and ULF fields (0–~20 Hz), and three searchcoil magnetometers to measure AC magnetic fields from around 10 Hz to 2 MHz. This implementation allows magnetic field coverage from the largest fields and lowest frequencies to the small plasma wave fields at high frequencies. Sensitivities range from 2 nT bit^{-1} at DC to greater than 3 fT bit^{-1} at 16 kHz. The combined magnetic fields instruments, including both fluxgate and searchcoil sensor sets, pigtails and analog board, weighs about 1.9 kg, and consumes about 1.5 W of power. The digitization of the output signals, and subsequent signal processing, are handled by a central FAST fields processor (see Ergun *et al.*, this volume), which makes sure that both E-field and B-field quantities are precisely timed and co-sampled to facilitate intercomparison. The instruments have been operating nominally on orbit since launch in August 1996.

Acknowledgements

This work was supported by NASA order number S-57795-F. We are grateful to the FAST team for comments and suggestions.

References

Boehm, M. H., Clemmons, J. H. Wahlund, J -E., Eriksson, A., Eliasson, L., Blomberg, L., Kintner, P., and Höfner, H: 1995, 'Observations of an Upward-directed Electron Beam with the Perpendicular Temperature of the Cold Ionosphere', *Geophys. Res. Lett.* **22**, 2103.

Borovsky, J. E.: 1993, 'Auroral arc Thicknesses as Predicted by Various Theories', *J. Geophys. Res.* **98**, 6101.

Clemmons, J. H., Boehm, M. H., Blomberg, L. G., Wahlund, J. -E., Eliasson, L., Lühr, H., and Haerendel, G.: 1995, 'Upwardly Accelerated Auroral Electrons: The Role of Parallel Electric Fields' (Abstract), *EOS Trans. Am. Geophys. Union* **76**, F551.

Erlandson, R. E., Zanetti, L. J., Acuña, N. H., Erksson, A. I., Elliasson, L., Boehm, M. H., and Blomberg, L. G.: 1994, 'Freja Observations of Electromagnetic Ion cyclotron ELF Waves and Transverse Oxygen Ion Acceleration on Auroral Field Lines', *Geophys. Res. Lett.* **21**, 1855.

Gurnett, D. A., Huff, R. L., Menietti, J. D., Winningham, J. D, Burch, J. L., and Shawhan, S. D.: 1984, 'Correlated and Low-Frequency Electric and Magnetic Noise Along the Auroral Field Lines', *J. Geophys. Res.* **89** 8971.

Marklund, G: 1997, 'Auroral Phenomena Related to Intense Electric Fields Observed by the Freja Satellite', *Plasma Phys. Control. Fusion* **39**, 195.

Marklund, G., Karlsson, T., and Clemmons, J.: 1997, 'On Low-Altitude Particle Acceleration and Intense Electric Fields and their Relationship to Black Aurora', *J. Geophys. Res.* **102**, 17 509.

Marklund, G., Blomberg, L., Fälthammar, C.-G., and Lindqvist, P. -A.: 1994, 'On Intense Diverging Electric Fields Asociated with Black Aurora', *Geophys. Res. Lett.* **21**, 1859.

Russell, C. T., Snare, R. C., Means, J. D., Pierce, R. D., Dearborn, D., Larson, M., Barr, G., and Le, G.: 1995, 'The GGS/Polar Magnetic Fields Investigation', *Space Sci. Rev.* **71**, 563.

Sugiura, M., Maynard, N. C., Farthing, W. H., Heppner, J. P., and Ledley, B. G.: 1982, 'Initial Results on the Correlation Between the Magnetic and Electric Fields Observed from the DE-2 Satellite in the Field-Aligned Current Regions', *Geophys. Res. Lett.* **9**, 985.

Weimer, D., Goertz, C. K., Gurnett, D. A., Maynard, N. C., and Burch, J. L.: 1985, 'Auroral Zone Electric Fields from DE 1 and DE 2 at Magnetic Conjunctions', *J. Geophys. Res.* **90**, 7479.

SCIENCE OPERATIONS AND DATA HANDLING FOR THE FAST SATELLITE

J. P McFADDEN, R. E. ERGUN, C. W. CARLSON, W. HERRICK, J. LORAN,
J. VERNETTI, W. TEITLER, K. BROMUND and T. QUINN
Space Sciences Laboratory, University of California, Berkeley, CA 94720, U.S.A.

Abstract. The University of California, Berkeley, is the primary science analysis center for the Fast Auroral Snapshot Explorer (FAST). It is responsible for processing and routing the science data to the various investigators and serves as an on-line data archive and science command center for the project. Raw data, received from NASA within a few hours of its transmission to the ground, is processed into CD-ROMs for archive and distribution to the Co-Investigators, and summary data is generated for viewing at the FAST Web site or for retrieval as key parameter data from NASA or the FAST Web site. Daily science command loads and real-time commands are generated to optimize science data collection. This paper presents a description of the FAST ground operations performed at Berkeley. Also included is a discussion of the data analysis software and tools that allow the large data volume (> 1 terabyte) to be accessed quickly and efficiently by the scientists. These tasks are more extensive than those performed in the past by science institutions and can be considered as a model for future small programs to reduce cost and maximize science return.

1. Introduction

The Fast Auroral Snapshot Explorer (FAST) was designed to investigate the plasma physics of auroral phenomena at extremely high time and spatial resolutions. It is the second of the Small Explorer (SMEX) Missions which were established to provide rapid, low cost opportunities to the space science community using a single Principal Investigator approach. The primary science objectives for this mission necessitated high data volume, real-time command capability, and control of science data collection on suborbital time scales. Since these requirements could increase costs to the project beyond the SMEX budget, the project adopted a nonstandard approach to its ground operations. This approach includes the use of semi-automated ground equipment to reduce personnel costs, and the offloading of specific tasks to the science institutions to shorten command turnaround time and reduce data collection costs. Descriptions of the NASA semi-automated ground stations, data packet processing equipment, and the Project Operation and Command Center (POCC) can be found in NASA/Goddard publications (NASA Document No. 822-DEM-1724, 1996; Shi *et al.*, 1993; SMEX FAST POCC System Users Guide, 1994). This paper presents a description of the FAST ground

Space Science Reviews **98:** 169–196, 2001.

operations performed at Berkeley, including a discussion of the commanding, data archiving, real-time software, and data analysis tools.

The Science Operation Center (SOC) at the University of California, Berkeley, is the primary science analysis center for the Fast Auroral Snapshot Explorer (FAST). The SOC is responsible for processing and routing the science data to the various investigators and serves as an on-line data archive and science command center for the project. The SOC receives level 0 (raw, time ordered) data from the NASA Goddard Space Flight Center within a few hours of transmission to the ground, processes the data into CD-ROMs for archive and distribution to the Co-Investigators, and generates summary data for viewing at the FAST Web site or for retrieval as key parameter data from the National Space Science Data Center or the FAST Web site. The FAST Key Parameters (KPs) are ingested into the NASA International Solar-Terrestrial Physics (ISTP) data base where they are included on both the ISTP KP CD-ROMs and on the ISTP CDAWeb site (http://CDAWeb.gsfc.nasa.gov). Daily science command loads and real-time commands are also generated at the SOC to optimize science data collection.

The paper is organized into four parts. It begins with a discussion of the science commanding, including real-time commanding and daily command loads performed by the SOC. The second section describes the data archive processing, including the generation of key parameter data. The third and fourth sections describe the real-time and data analysis tools developed for the scientists. It is hoped that this paper provides not only an overview of the FAST specific ground operations and software, but also a general outline of SOC tasks and science software design which can be used as a model for future small programs to reduce cost and maximize science return.

2. Science Commanding

The FAST spacecraft requires that daily commands be loaded into the stored command memory buffer for normal operations. These command loads include spacecraft commands generated by NASA such as attitude control system commands to preserve the nominal reverse cartwheel attitude, transmitter on/off commands, and telemetry channel enable/disable commands. Science commanding consists primarily of science instrument mode commands, trigger commands, and memory management commands. Due to power and data volume constraints imposed on the FAST spacecraft, data collection and instrument power are cycled. This cycling is performed by the science mode commands which selectively turn on and off the instruments during the orbit. These stored science commands are limited to 110 per day (10 per orbit).

2.1. SCIENCE MODE COMMANDS

Science mode commands are single commands that trigger execution of a command set stored on the spacecraft that completely configures the instruments. Science modes are specified by two one-byte codes that specify the field and particle experiments separately. Eight particle and eight field modes are stored onboard at any one time allowing up to 64 combinations. Although real-time mode commanding can be used, selection of instrument modes is generally performed on the ground where they are included in the daily stored command load. As new modes are developed, they are uploaded, replacing the older modes. Mode status is tracked on the ground to assure that a discontinued mode is not called out in any command set.

Instrument modes are used to configure the instruments to emphasize a particular science observation or to select the data collection rates to match data volume constraints. Particle modes are used to configure the 2 ion and 14 electron electrostatic analyzers (Carlson *et al.*, 2001 this issue) and the TEAMS mass spectrometer (Klumpar *et al.*, 2001, this issue). The mode sets the instrument sample rate (1.6–6.5 ms), high voltage sweep rate (64–16 sweeps spin^{-1}) and energy range (typically 2 eV to 12 keV for the mass spectrometer, 3 eV to 25 keV for the ion spectrometer, 4 eV to 30 keV for the electron spectrometer, and 6 fixed energies between 64 eV and 16 keV for the high time resolution electron spectrograph). Field modes determine electric (Ergun *et al.*, 2001, this issue) and magnetic sensor (Elphic *et al.*, 2001, this issue) measurement combinations. There are 10 electric field and 6 magnetic field sensor measurements which can be combined or multiplexed in various ways for the low rate survey data, the swept frequency analyzers, 8 high rate burst A/D channels, and 4 high frequency waveform capture A/D channels. The field mode also determines filtering, selection of electric field probe configuration (voltage or current), and sample rates (8 Hz to 2 MHz sampling). Details of the more than 100 FAST particle and field modes can be found on the FAST web site (http://sprg.ssl.berkeley.edu/fast). The mode also determines four data rates that are part of the FAST science optimization scheme. These four rates are referred to as Slow Survey, Fast Survey, Burst, and High Speed Burst Memory (HSBM). Slow survey is the slowest data collection rate and these data are used to provide a large-scale context for the higher resolution measurements. Slow survey data collection rates are comparable to previous satellite measurements with particle and wave spectra gathered on spacecraft spin time scales, and with low frequency waveforms having 30 ms resolution. Fast survey data collection is typically 8 to 32 times the slow survey rate and is triggered onboard in active regions, replacing slow survey data collection. Fast survey rates (230–1100 kbit) can be comparable to the highest burst memory collection rates on previous satellites. During orbits having consecutive memory dumps, fast survey data collection can be maintained through most of the auroral zone crossings.

Burst data collection is reserved for the most interesting events and is triggered by onboard algorithms that sample a variety of field and plasma parameters. Waveforms are sampled at 32 kHz, high frequency wave spectra are generated at 16 Hz, high resolution (48 energy × 32 angle) electron and ion distributions are measured with 78 ms resolution, and coarse electron distributions (6 energy × 16 angle) are measured with 1.6 ms resolution. These data are typically 1–2 orders of magnitude higher in time resolution than the fast survey data, thus they quickly fill the available onboard memory. Onboard trigger and event capture evaluation algorithms are used to identify the best events for selective transmission to the ground. Finally the HSBM data capture is used to measure high frequency waveform (2 MHz sampling) and also includes a separate trigger and event evaluation algorithm.

2.2. Data Collection Triggers

In addition to the instrument mode commands, trigger commands are also sent as part of the daily science stored command loads. Eight trigger algorithms are stored onboard providing a selection of criteria for optimizing both fast survey and burst capture. Table I lists the inputs to the trigger algorithms and includes five current non-trivial algorithms. The simplest algorithms use the raw inputs such as the total low frequency filtered power (Ch4 or Ch5) which triggers on lower hybrid and whistler waves, the DC (<10 Hz) electric field magnitude (Ch0 or Ch1) that triggers on electrostatic shocks, or electron energy flux (Ch22 to Ch29) which triggers in enhanced auroral precipitation. More sophisticated algorithms that involve two or more inputs can be tailored to provide more reliable triggers for specific science (bottom of Table I). For example an algorithm that uses the density (Ch3) and high frequency wave power (Ch6) is an excellent trigger for selecting events with Auroral Kilometric Radiation and the associated density cavities and ion beams. The stored command loads allow selection of various triggers during an orbit or day to emphasize different science.

Figure 1 shows an example of FAST data gathered at different rates and the usefulness of the triggering algorithms. The top four panels show the electric and magnetic fields, electrons, and ions across a portion of the auroral oval. Although difficult to discern from the figure, a transition from slow survey (48 kbit s^{-1}) to fast survey (432 kbit s^{-1}) data rate occurred at 20:48:24 as FAST entered the auroral electron precipitation region, and a transition back to slow survey rate occurred at 20:55:00 as the spacecraft left the nightside aurora and entered the polar cap. These data rate transitions were triggered by the onboard algorithm that measured the electron energy flux. The second transition occurred about 30 s after entering the polar cap due to onboard filtering which prevents rapid toggling between data collection rates. The middle three panels show burst data captured by the electrostatic shock trigger algorithm. The large electric fields (panels 1 and 5) that initially triggered the burst are associated with intense upgoing electron beams (panel 6) and heated ions (panel 7) in the downward current region (positive

Table I

List the inputs to the trigger algorithms and 5 typical multi-input algorithms

Electric field inputs

Channel	Quantity	Frequency	Science trigger
0	V1-V4 DC	< 10 Hz	Electrostatic Shocks
1	V5-V8 DC	< 10 Hz	Electrostatic Shocks
2	ELF V1-V4	10 Hz - 1 kHz	Ion Cyclotron waves
3	Ne_6	< 10 Hz	Density cavities
4	LFF V1-V4	1 kHz - 16 kHz	Whistlers, Lower Hybrid
5	LFF V5-V8	1 kHz - 16 kHz	Whistlers, Lower Hybrid
6	BBF V1-V2	>200 kHz	AKR, Langmuir
7	BBF V3-V4	>200 kHz	AKR, Langmuir

Particle inputs

Channel	Quantity	Science trigger
8–9	i+ ESA1/2 max counts	Ion beams, conics
10–11	e- ESA 3/4 max counts	Arcs, Upgoing e-
12	i+ ESA 1/2 peak energy	Ion beams, conics
13	e- ESA 3/4 peak energy	Arcs, Upgoing e-
14–17	i+ ESA1 Energy Flux, 4/sweep	Ion beams, conics
18–21	i+ ESA2 Energy Flux, 4/sweep	Ion beams, conics
22–25	e- ESA3 Energy Flux, 4/sweep	Arcs, Upgoing e-
26–29	e- ESA4 Energy Flux, 4/sweep	Arcs, Upgoing e-

Typical algorithms

Algorithm	Science trigger
AKR = Ch6 + (255- Ch3)/2	AKR in density cavity
ESS = abs(Ch1 − 128) × 2	Electrostatic shocks
VLF = (Ch5 + Ch4 + 255 − Ch3)/4	VLF in low density
MAX = Max(AKR, ESS × 10, VLF)	General field trigger
EF = (Ch24 + Ch28)/2	e- Energy Flux, >500 eV

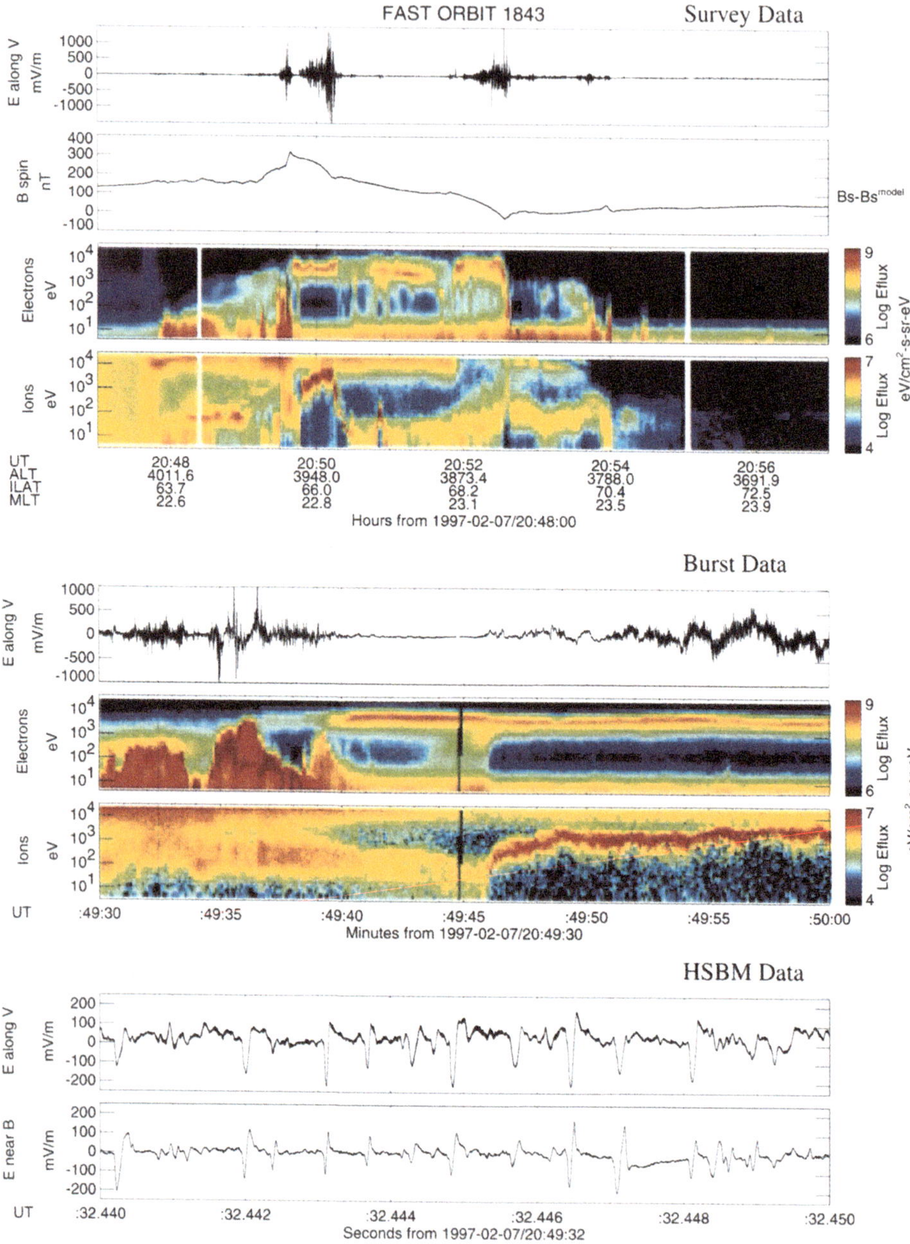

Figure 1. FAST data captured during a nightside auroral crossing. The top four panels show field and particle data collected at survey data rates. The middle three panels show burst data whose collection was triggered by an onboard algorithm sensitive to large electric fields. The bottom panels show high speed burst memory waveform capture of fast electron solitary waves that are common in downward current regions.

slope in the spin axis or eastward magnetic field component in panel 2). These downward current regions have been shown to contain fast electron solitary waves (or electron holes) that travel up the field line at the electron drift speed (Ergun *et al.*, 1998). The bottom two panels show the HSBM capture of the solitary waves showing a unipolar perpendicular (panel 8) and bipolar parallel (panel 9) electric field signal. The burst capture continued to trigger during the large electric fields associated with the upgoing ion beam spanning 20:49:46 to 20:50:18. Burst data (not shown) was also captured during the large electric fields at 20:52:30 amid intense electromagnetic ion cyclotron waves and at 20:53:42 during a second up-going electron beam. The multiple data rates and onboard triggers have allowed the FAST spacecraft to capture scientifically interesting data on nearly every orbit.

2.3. Data Management

Since the majority of data is generated by the instruments, the FAST solid state data recorder (125 MBytes) was incorporated into the science electronics package. To accommodate the various data products, the FAST data recorder is segmented into five partitions. These partitions contain packet headers, housekeeping, survey data, burst data, and HSBM data. To simplify data collection and readout, a standard 1024 byte packet is used for science data, with 40 bytes of packet header stored separately. Housekeeping packetization is handled by the spacecraft electronics developed by NASA, but housekeeping data is stored in the mass memory. The housekeeping data partition is treated as a FIFO (First-In-First-Out) but takes multiple (>6) orbits before it rolls over due to low housekeeping data rates, assuring that data is not lost even during days with only a few contacts. The burst and HSBM memory segments use selection criteria to overwrite events with poorer evaluation thus preventing memory overflow. The survey memory segment, however, is a FIFO and the high fast survey data rate may cause it to overflow.

To prevent survey memory overflow, ground software is used to manage this memory partition with a 'Fast Survey Allocation Command' which limits fast survey data collection. Onboard data transmission priority assures that the survey memory partition is emptied at the end of each nominal contact. Ground software is used to calculate the fraction of fast survey data collection that can be sustained without overflowing the memory before the next contact. The program uses as inputs the contact schedule provided by NASA, the mode data rates, and the data collection times. The Fast Survey Allocation Commands are added to the stored command load and executed following each contact.

2.4. Power Management

In order to minimize spacecraft weight and achieve a higher apogee, the FAST power system was matched to science requirements and does not support continuous operation of all onboard systems throughout the mission. Figure 2 shows the prelaunch estimate of solar array power during the first six months of the mission.

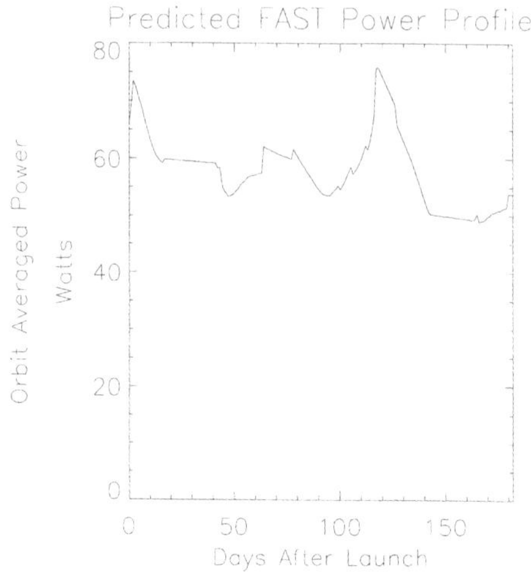

Figure 2. The calculated solar array power during the first 6 months of the mission. The average spacecraft power is about 70 watts which exceeds the orbit averaged solar array power during most of the mission. Instead of using a larger solar array which would have added mass and reduced the FAST apogee, the scientists chose to power cycle the experiments. The instruments are turned on for about 30%–40% of the orbit, during auroral zone crossings, reducing the orbit averaged power to about 50 watts.

Variations in the curve are due to changes in the spacecraft spin vector with respect to the Sun and due to periods of eclipse as the orbit precesses rapidly about the Earth. Available power will also degrade with time due to radiation damage.

The spacecraft power budget for all the systems is shown in Table II. With all systems powered and assuming a nominal 30-min transmitter contact and 10 min of magnetic torquing for attitude control, the spacecraft requires about 70 W. This level is well above the available orbit averaged power for much of the mission, necessitating power cycling of the experiments.

Power cycling is achieved by using stored command loads to tell the experiments when to turn on/off. The primary purpose of these commands is to limit data collection to those periods when auroral science can be optimized, namely the high altitude auroral zone crossings. However, a secondary purpose essential to the mission is to limit power consumption by the experiments. This is achieved by using a low power 'back-orbit' mode for about $\frac{2}{3}$ of the orbit. The back-orbit mode requires only 7.7 W from the experiments, as opposed to 37.1 W for normal science modes. During back-orbit mode only low rate magnetic field data is collected and most experiments are turned off. Normal science modes are typically selected only for the auroral crossings in the Earth hemisphere containing apogee.

2.5. Sun–Nadir Table Loads

In addition to the stored command loads, ground software is also required to generate a Sun–Nadir Table of angles. Particle data on FAST is sorted by spin phase

Table II

The spacecraft subsystem power budget

Spacecraft subsystem	Nominal power	Back orbit* power	Duty cycle	Average power
Mag S/C	0.5		1.0	0.5
Sun sensor	0.6		1.0	0.6
HCI	0.7		1.0	0.7
Spin coil	5.1		0.**	0
Precession coil	4.3		0.063***	0.27
MUE	13.5		1.0	13.5
Shunt drivers	0.5		1.0	0.5
Receiver	4.5		1.0	4.5
Transmitter	28.3		0.23****	6.5
Battery heaters	1.0		1.0	1.0
Other heaters	4.0		1.0	4.0
IDPU	18.3	6.6	0.3	10.11
E-field booms	2.0	0	0.3	0.60
Mag flux gate	0.08	0.08	0.3	0.08
Mag search coil	0.50	0	0.3	0.15
ESA stacks	11.75	0	0.3	3.53
TEAMS	4.5	1.0	0.3	2.05
	–			–
	100.13 W			48.59

*Backorbit instrument power is defined as non-science data collection modes.

**Spin coils are used primarily during early orbit operations to spin up spacecraft during boom deployments.

***Precession coil is typically run for 45 min. twice a day during Backorbit.

****Assuming one 30 min contact.

relative to the Sun. In order to accurately maintain this spin phase sorting during eclipse, onboard algorithms were developed to calculate the sun spin phase pulse based upon horizon sensor pulses, the midpoint of which give the Earth nadir direction. In order to calculate the sun pulse, a table of Sun-to-Nadir angles and delta angles, where delta angles are the change in sun-to-nadir angle per orbit, are uploaded daily. Due to the highly elliptical orbit and rapid precession rates, the table is non-trivial and changes significantly over a period of a few days. This Sun-to-Nadir Table allows the calculation of the sun-to-nadir angle, projected onto the spacecraft spin plane, as a function of time. Use of the onboard algorithm provides a smooth transition for the particle sorting when entering or leaving eclipse.

2.6. Command Generation Software

The mode, trigger, and fast survey allocation commands and the Sun-to-Nadir tables are generated for the daily command load by an automated program. This program takes as inputs a contact schedule (weekly file), a predicted orbit vector (daily file), a predicted spacecraft spin vector (weekly file), and a power budget (weekly file) all provided by NASA. The program also uses a file of rules generated at the SOC to customize the commands. The rules file allows specification of the spacecraft mode and trigger as a function of invariant latitude, magnetic local time and altitude. These parameters allow some flexibility in the automated mode selection. For example, different modes for midnight and cusp auroral crossings can be used to optimize different science in these regions.

At the start of a command generation session, the orbit vector is propagated out for three days and used to build three days of Sun-to-Nadir tables and stored command loads. Three days of commands and tables are generated and sent to the POCC to provide continuity in case network lines go down. During the second year of the mission, a five-day work week was implemented to reduce mission costs. To accommodate this change, the daily (24-hour) spacecraft command loads were replaced by 48-hour loads.

The automated command generation program works as follows. Data collection periods, and their associated Instrument Mode Commands, are determined from the calculated orbit and the rules file. Typical data collection is limited to periods above 60 deg invariant latitude in the hemisphere containing apogee. Once data collection periods are determined, the power is calculated for each orbit and compared to the available power budget. If the calculated power exceeds the available power, the data collection period is reduced appropriately. Burst Trigger and Fast Survey Trigger Commands are also generated and inserted in the load. Next, the Fast Survey Allocation Commands are calculated, based upon contact schedule and instrument mode data rates, and inserted into the stored command load. Sun-Nadir table loads are also automatically calculated and included. At this point, an opportunity is given for a SOC operator to check the loads before they are sent to NASA. The SOC operator also has the option of editing the command load, which may require a recalculation of the power allocation and Fast Survey Allocation Commands. The automated command generation program provides a convenient commanding method to generate daily command loads, and includes the flexibility needed to allow commands to be tailored to the orbit in order to accommodate special events such as multiple spacecraft conjunctions or coordination with ground-based operations.

2.7. Real Time Commanding

Although stored command loads are the primary means of controlling the FAST experiment, real-time commanding is available through the FAST Project Operation and Command Center. Individual commands or batch command procedures can be initiated with a phone call. Commands are checked for validity and format with the FAST command database before transmission. As new commands are developed, they can be directly entered into the command database from Berkeley. Turn around time for the development of a new command procedure, incorporation into the database, and execution can be reduced to a few minutes, allowing changes in instrument operation to be initiated during a single orbit. During the FAST campaign, turn around time for commanding was less than 1 min and allowed rapid changes in instrument configuration based upon ground observations or real-time science analysis.

3. FAST Data Archive

The FAST SOC is the primary data collection and data archive processing facility for the project. The FAST science team took on this task, normally provided by NASA, to reduce costs to the project and to create an on-line data center that could handle the large volume of data (>1 terabyte). The large data volume is inherent to the mission, which specifies the collection of high rate measurements in the auroral regions each orbit. The 125 MByte solid state memory on the FAST satellite, together with a contact nearly each orbit (133 min orbital period), has produced a data volume approaching a terabyte during the first 2 years. In order to handle the data at a low cost, an automated archive system was developed that requires minimal personnel.

3.1. Data Collection

The data path for FAST begins on the spacecraft where the onboard solid state memory is used to collect high rate data and buffer it before transmission to the ground. Some of these data are preselected by onboard trigger and 'goodness' factor algorithms as higher quality data with priority for transmission to the ground. A table of data priorities and data rates is shown in Table III. Onboard software selects the data with the highest priority for first transmission, but limits the data rate for each quantity to its maximum rate. In general all data are sent to the ground and the memory emptied for nominal contacts. However, on those contacts where data collection is simultaneous with transmission, not all the Burst and HSBM data (VC 2) will be transmitted since these data are filled faster than can be read out. Here the selection criteria, based upon the goodness of each burst, is used to select priority for burst data readout. For more information on data selection, see Harvey *et al.* (2001, this issue).

Data are received by a number of ground stations. The primary ground stations are a Transportable Orbital Tracking Station (TOTS, see NASA Document No. 822-DEM-1724, 1996) located at Poker Flat, Alaska, and the NASA Wallops Flight Facility located at Wallops Island, Virginia. NASA Deep Space Network stations

Table III

FAST virtual channel encoding including data priorities and data rates

Virtual channel	Data type	Data priority	Bandwidth at 900 kHz	Bandwidth at 1500 kHz	Bandwith at 2250 kHz
VC0	Real time housekeeping	1	Unlimited	Unlimited	Unlimited
VC1	Stored housekeeping	2	49.2	500	563
VC2	Burst data	6	Unlimited	Unlimited	Unlimited
VC2	HSBM data	7	Unlimited	Unlimited	Unlimited
VC3	Rate limited survey data	4	49.2	49.2	49.2
VC4	Nominal survey data	5	900	1000	1000
VC5	Engineering data	3	Unlimited	Unlimited	Unlimited
VC7	Fill data	8			

at Canberra, Australia and Santiago, Chile are also used. An ESA Station near Kiruna, Sweden was used during the FAST 1997 ground coordinated observing campaign, and a new TOTS at McMurdo, Antarctica, became available in mid-1997. The primary ground stations send the data, rate buffered in real-time, over dedicated lines to the Packet Processing System (PPS, see Shi *et al.*, 1993) located at the Goddard Space Flight Center. Other stations record data on tape and mail the tapes back to the PPS.

The PPS is a semi-automated system developed to handle data for the FAST project. The PPS repackets and stores the telemetry as it is received. Once the contact is completed, it sorts the data by data type into separate files, ordering it in time, eliminating duplications, and removing untrustworthy packets. The resulting data set is transmitted to secure computers in the SOC over a dedicated network. Once the data has arrived at the SOC, it is checked for size and the result posted on the FAST Web page. This page is viewed by PPS personnel for comparison with their records to assure that the SOC receives all the data.

3.2. Data Processing and Archive

Figure 3 shows a block diagram of the automated processing that is triggered once the data is received at the SOC. During this processing, the data is held on the secure network until enough data has been received to fill a CD-ROM (650 MBytes). A disk image is then built and transferred to an open data analysis net-

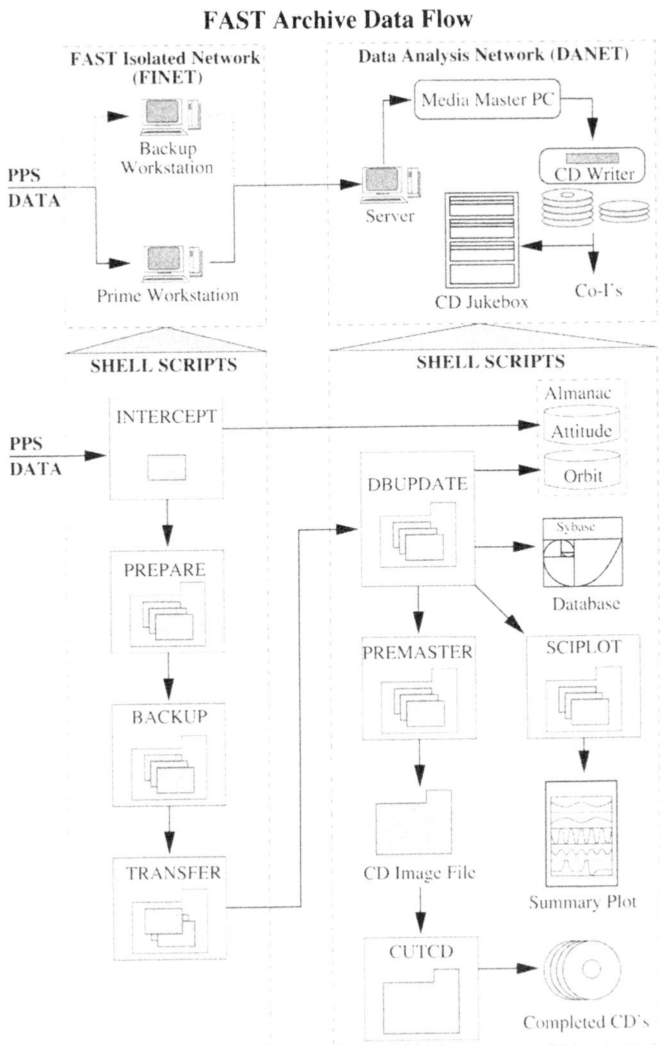

Figure 3. A block diagram of the data archive process. Scripts are run on both the Fast Isolated Network (FINET) and on the Data Analysis Network (DANET) to complete the archive. The INTERCEPT script senses a new data set has been received and processes the data generating a 'fastcount file' that is stored with the raw data. The fastcount file is used by data analysis tools to determine memory (RAM) allocation requirements for analyzing the data and by the database update (DBUPDATE) script. In addition, fastcount information is posted on the FAST Web page for internal use and to allow PPS personnel to check data volume received. INTERCEPT launches the PREPARE script which determines if the volume of data currently on FINET exceeds that of a CD-ROM (650 MBytes). INTERCEPT performs further processing as well as launching the BACKUP and TRANSFER scripts when a CD-ROM data set is ready. The PREPARE scripts generates a UNIX directory structure that will mirror the CD-ROM. BACKUP copies across the 'firewall' software that isolates the FINET and DANET networks. Automated scripts are run on DANET to complete the data archive once each CD-ROM file set is received. DBUPDATE updates the FAST database and the orbit-attitude almanac files. The PREMASTER script forms an 'image' (single file) of the cd. The SCIPLOT script generates summary plots of the data. The CUTCD script transfers the CD 'image' to a Young Minds Media Master which directs a CD writer to make the CD-ROMs.

work where the CD-ROMs are actually created. Typical contacts have data volumes that range from 50 to 300 MBytes, so multiple data sets are required to fill a CD-ROM. Routines are run on both of these networks in support of the archive process (see Figure 3). In the final step, a Young Minds Media Master (Young Minds Inc., Redlands, CA, http://www.ymi.com) directs a CD writer (Kodak CD Writer 600) to make the CD set (ISO 9660 with Rock Ridge Interchange and System Use Sharing Protocols that provides UNIX-style file and directory names with complete ISO 9660 compatibility). The CD writer includes robotics that hold 75 blank CDs and automatically loads and unloads CDs during the processing, and an image printer which labels the CDs after they are written. Eight CDs are created for distribution to the CoIs, to the NASA archive, and for loading into the on-line data archive at Berkeley.

The on-line data archive consists of an array of jukeboxes (Kubic model CDR-240) attached to a computer on the open data analysis network. Each jukebox holds 240 CDs and contains 4 CD readers which can be independently addressed by the jukebox software. The entire system of jukeboxes is controlled by a software package provided by Luminex (Fire Series Software, Luminex Software Inc., Riverside, CA., http://home.luminex.com), which runs the internal robotics, controls SCSI communications and handles disk access contention, making the system appear as a UNIX file system. CDs are entered or removed from the jukeboxes by programs developed at Berkeley that interface to the jukebox software. The jukeboxes are rather compact (36 in $\times$ 17 in $\times$ 7 in) and the entire on-line archive fits in two standard 19 inch racks.

Additional processes are run on the open network to complete the data archive once each CD-ROM file set is received. The primary task is to update the FAST database. The large volume of data files, over 1 million files organized by contact and data type, require a sophisticated database to keep track of files and to allow selection of data based on criteria other than time. A commercial product (Sybase, Sybase Inc., Emeryville, CA, http://www.powersoft.com) was chosen and a baseline set of tables were created to organize the database. Figure 4 shows the table structure and elements of each table. In addition to the tables containing file information, the database contains an events_data table with ephemeris, attitude, and model magnetic field. Events_data entries exist for all operational_events (science mode changes) and ephemeris_events (orbital highlights) entries. An operational_events entry is also generated every 20 s to assure minimal resolution in the database for searches based upon ephemeris. Science tables will be added to the database as they are developed to allow searches based upon auroral characteristics, such as energy flux, maximum wave power, and convection flow.

3.3. Summary Data

The data archive processing also includes the production of summary plots and key parameter data sets. The summary plot routine determines which orbits are

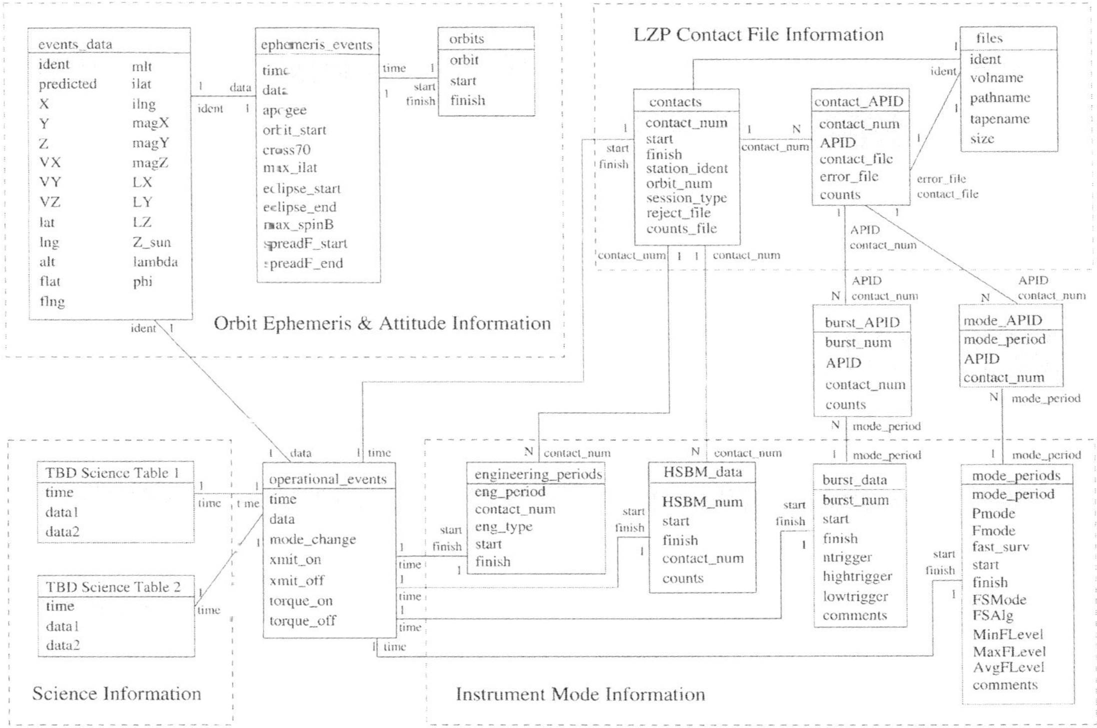

FAST Database Tables

Figure 4. FAST database table structure and elements of each table. Sybase is used as the database platform and the database resides at Berkeley. Database tables are automatically updated as part of the data archive process. Scientists access the database primarily through queries generated by the data manager program. This program can be run locally or from a remote site to obtain a list of data files based upon data type, time, ephemeris, or spacecraft operating mode. The data manager program also interfaces to the data analysis program, automatically generating queries based upon the selected plot types, and can be used to download files from the on-line data archive at Berkeley. The database can also be accessed directly using the WISQL, a convenient user interface for entering database queries.

currently present in the CD data set, then launches a batch process that generates summary plots and CDF (Common Data Format, http://nssdc.gsfc.nasa.gov/cdf) files for each orbit. The batch process uses both the FAST data analysis program (SDT, Science Data Tool, developed by the FAST science team and discussed below) and IDL (Interactive Data Language, Research Systems Inc., Boulder, CO, http://www.rsinc.com/idl) to generate 1 spin average data sets for five summary plot data types: Ions, Electrons, Mass Spectrometer, AC Fields, and DC Fields. GIF plots are made and placed immediately on the FAST Web site for viewing. Plots cover 20-min periods prior and post the highest invariant latitude for each polar crossing during science data collection, so typically two or four GIF files are generated for each data type each orbit. In addition, 5 CDF files are generated containing summary data for each plot type, which can be downloaded from the FAST Web site. A description of the summary plot and CDF file contents is given in Table IV.

Key parameter data sets are produced for inclusion in the ISTP Key Parameter (KP) data base from the CDF files described above. Key parameter files are made by combining all (typically 11) of the single orbit CDF files of each data type for a day into one large file. Most collaborative studies, such as spacecraft or ground conjunctions, require only a single orbit to be downloaded over the net. Thus scientists may prefer to use the FAST Web page as a faster means of obtaining a single orbit's data. For those interested in obtaining complete key parameter data sets, we recommend using the NASA centers. ISTP KP values in CDF format can be accessed through the ISTP CDHF (http://www-istp.gsfc.nasa.gov/istp/science/kpi) and the NSSDC Spycat (http://nssdc.gsfc.nasa.gov/space/ndads/spycat.html) and CDAWeb (http://CDAWeb.gsfc.nasa.gov) systems for on-line retrieval and plotting.

3.4. Ephemeris and Attitude Data

A final task of the data archiving procedures is the production of ephemeris and attitude data sets. To handle ephemeris data, an orbit propagator and field line tracing program was developed that uses a NASA supplied orbit propagator, a field line tracing program using the IGRF 1995 magnetic field model with secular variation (http://www.ngdc.noaa.gov/IAGA/wg8/wg8.html), and takes an orbit vector supplied by NASA (one per day) as an input. The program generates spacecraft GEI (Geocentric Earth Inertial) position and velocity, latitude and longitude, invariant latitude, Magnetic Local Time (MLT), and latitude and longitude of the magnetic footprint (at 100 km altitude). Rather than store ephemeris data with time resolution that allows linear interpolation, the FAST team decided to rely on a small ephemeris file which contains one vector (position, velocity in GEI coordinates) for each orbit. The orbit propagator program is then used to quickly calculate ephemeris data from the nearest time in the file.

Similarly, an attitude propagator program was developed to calculate spacecraft attitude from the spacecraft attitude housekeeping data and magnetometer data. An

Table IV

FAST summary plot and CDF file contents

EES – Electron Electrostastic analyzer Survey data
Electron Energy Spectra, 0–30 deg pitch angle
Electron Energy Spectra, 60–120 deg pitch angle
Electron Energy Spectra, 150–180 deg pitch angle
Electron Pitch Angle Distribution, 0.1–1.0 keV
Electron Pitch Angle Distribution, 1.0–30. keV
Electron Energy Flux, mapped along B to 100 km altitude
Electron Number Flux, mapped along B to 100 km altitude
IES – Ion Electrostatic analyzer Survey data
Ion Energy Spectra, 0–30 deg pitch angle
Ion Energy Spectra, 40–140 deg pitch angle
Ion Energy Spectra, 150–180 deg pitch angle
Ion Pitch Angle Distribution, 0.05–1.0 keV
Ion Pitch Angle Distribution, 1.0–25 keV
Ion Energy Flux, mapped along B to 100 km altitude
Ion Number Flux, mapped along B to 100 km altitude
TMS – Time-of-flight Mass spectrometer Survey data
Hydrogen Energy Spectra 0–360 deg pitch angle
Helium Energy Spectra, 0–360 deg pitch angle
Oxygen Energy Spectra, 0–360 deg pitch angle
Hydrogen Pitch Angle Distribution, 0.01–1.0 keV
Hydrogen Pitch Angle Distribution, 1.0–15 keV
Helium Pitch Angle Distribution, 0.01–1.0 keV
Helium Pitch Angle Distribution, 1.0–15 keV
Oxygen Pitch Angle Distribution, 0.01–1.0 keV
Oxygen Pitch Angle Distribution, 1.0–15 keV
ACF – AC Fields Survey data
HF Electric Field Spectra (0–2 MHz)
VLF Electric Field Spectra (0–16 kHz)
ELF Electric Field Spectra (0–2 kHz)
HF Magnetic Field Spectra (0–2 MHz)
VLF Magnetic Field Spectra (0–16 kHz)
ELF Magnetic Field Spectra (0–2 kHz)
HF Band Integrated Power
VLF Band Integrated Power
ELF Band Integrated Power
Spacecraft mode (slow survey, fast survey, burst)
DCF – DC Fields Survey Data
Electric field perpendicular to spin plane projection of B, roughly equatorward
Electric field along spin plane projection of B
Langmuir probe current
Delta B perpendicular to spin plane projection of B, roughly equatorward
Delta B $\pm$ spin axis, roughly West
Delta B along spin plane projection of B
Spacecraft potential
Angle between spin plane and B

attitude almanac file is generated using multiple orbits. Later calls for attitude data use an attitude propagator to calculate the attitude from the nearest data point in the almanac file. The FAST science team took on the task of determining the spacecraft attitude, normally performed by NASA, in order to reduce project costs and since accurate attitude determination is required for precise magnetic field solutions.

4. Real Time Software

One of the central objectives of the FAST program was the inclusion of a 'campaign' in which real-time data analysis would be used to coordinate spacecraft operations with ground based observations. The need for a real-time data analysis tool precluded the typical development path where instrument ground station equipment (IGSE) used for prelaunch testing and commanding is separate from the data analysis programs. Instead a series of programs that run on the data analysis platform (Sun SPARC stations running the Solaris 2 operating system) were developed to interface with the spacecraft GSE and the data analysis tools, producing a single combined IGSE analysis package.

Since the data analysis platform was UNIX based, it was not configurable as an event driven platform such as DOS based PCs. Instead, networking techniques were used to connect to the spacecraft GSE and draw data to the data analysis platform. A feature of the Solaris 2 operating system (threads: coordinated, cooperating subprograms) was used to duplicate the fast response inherent in event driven systems. In addition, the same protocol for communication between the spacecraft GSE and the IGSE were adopted for the other ground systems (TOTS and PPS), providing a single interface for all real-time science operations. This software package allows the scientist to monitor spacecraft operations and display science quantities simultaneously.

4.1. IGSE to GSE interface program

A block diagram of the real-time programs is shown in Figure 5. The interface to the spacecraft GSE (or TOTS or PPS) is provided by a relay program. This program has the following functions:

- Real-time data input from the spacecraft GSE, TOTS, or PPS.
- Playback data from a disk file created in a previous session.
- Data capture to a disk file.
- Data relay to monitor workstations on a secondary network.
- Data relay to a remote workstation on a wide area network such as the internet.
- Data set monitoring including data volume, data rate, and data packet statistics.

Each of these functions are implemented as separate threads (subprograms) in the relay software. The actual data resides in a systemwide shared memory area,

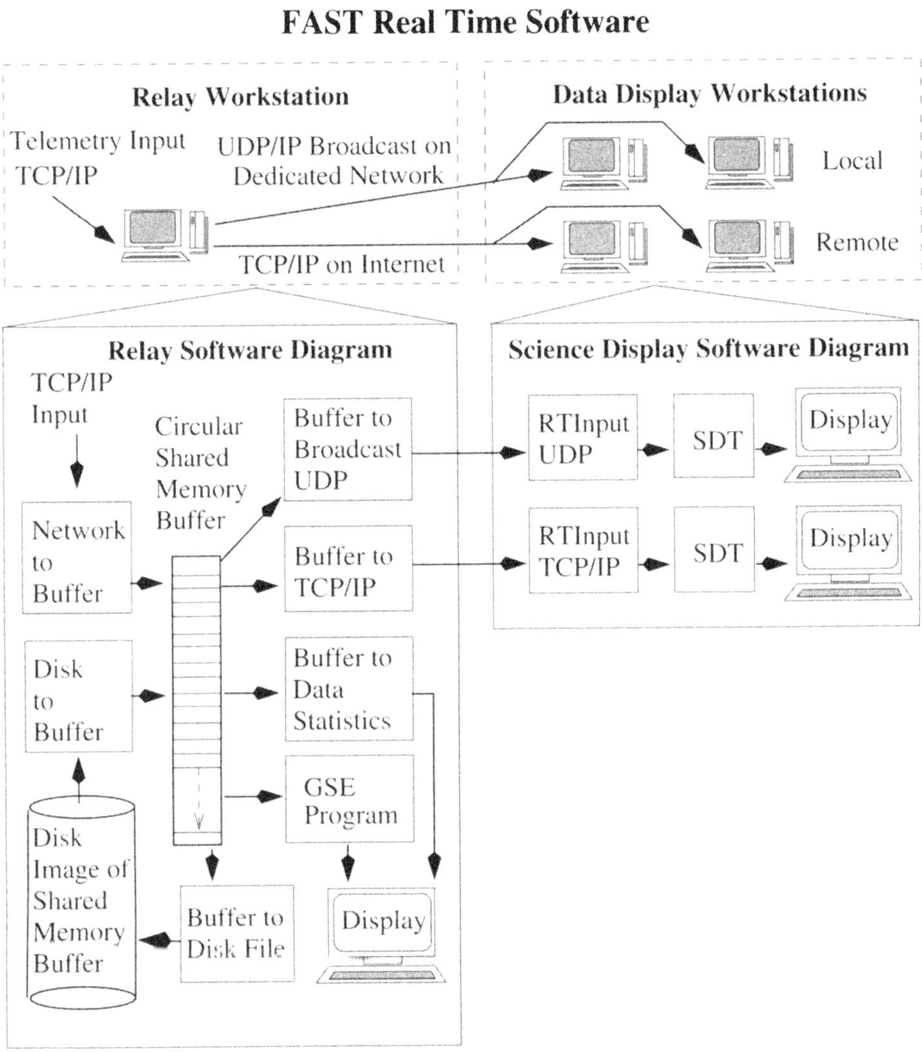

Figure 5. A block diagram of the real-time data capture and display programs. The relay workstation gathers telemetry from a NASA ground station into a circular memory buffer where the data can be copied to disk, broadcast to workstations on a local network, or transmitted over the internet to a remote workstation. Data statistics are displayed for the various data types, and a GSE program is usually run to display instrument housekeeping. Workstations attached to a local network or at remote sites can use the RTInput program to access selected subsets of the data for display using the FAST data analysis program, SDT.

which enables other programs to access the relay data flow. The data memory subsystem monitors how many threads are active, and when the last thread exits the shared memory buffer is removed.

The implementation of the shared memory subsystem is optimized for performance. All data flow functions are concurrent and independent, allowing the various computer subsystems, such as disk, network I/O and video to update in parallel. Access to the shared memory is not computationally intensive: threads reading the memory only acquire read locks on a single data block (approximately 1/1000 of the total shared memory) at one time, and the number of concurrent read locks on a given block are unlimited in number. Only threads writing to the memory acquire exclusive locks, and again only on one block at one time. A simple conditional locking scheme handles thread positioning and data availability within the data buffer.

A simple one character command interface is employed for easy control of all of the relay program's functions, and the display works entirely within a UNIX terminal window, so that the program can be efficiently executed from a remote system. The screen update rate can also be controlled via a start-up option so that slow network links to a remote display can work properly. The ability to remotely monitor the instruments prevented key personnel from being tied up for long periods at remote sites during instrument integration.

4.2. Housekeeping Display and Remote Monitoring

One program designed to read the memory buffers of the relay program is the FAST GSE program which displays housekeeping data, thereby allowing real-time monitoring of instrument health. This multi-threaded program reads data packets containing housekeeping data and generates a multipanel display which can be configured to display different spacecraft health and safety data. The user interface allows the scientists to change individual panels to check on specific instruments, or choose from pre-configured combinations of displays geared for specific tests. The GSE program is most useful for monitoring instrument health and safety during instrument integration, launch and early orbit, and can also be used to diagnose instrument problems by replaying telemetry from a disk file.

The broadcast feature of the relay program allows other nearby computers to listen in on the captured data stream. In this way each individual computer need only decode and display a limited portion of the data stream. Computers connect to the broadcast network using a data capture program (RTInput, Figure 5) that opens a UDP (User Datagram Protocol) connection with the relay program. The relay program broadcasts different science or housekeeping data types on individual network channels, so that the other computers can receive a limited data set by listening on the corresponding set of channels.

The data capture program reads packets from the UDP connection and writes them into its own shared memory buffer with the same format as is written by the relay program. Hence other programs, like the GSE program, can read these data as it would on the relay machine. The data capture program allows the user to select which data to receive from the data stream, and is usually used in conjunction with the FAST science display program, SDT. When a scientist selects data to be displayed in SDT, SDT automatically sends the data capture program a list of the data types which are used to configure the UDP connection. If a computer cannot keep up with the data it is trying to display, then UDP packets can be dropped without affecting the ability of any other computer to receive data.

The data capture program (RTInput, Figure 5) can also be used to store data to a disk file, to read a disk file, or to connect to a remote site over the internet. The disk read feature can be used to replay telemetry in near real-time, and is useful for diagnosing instrument performance or health and safety in conjunction with GSE and/or SDT programs. The scientist can select playback speed, pause, or step through the data forwards or backwards one telemetry packet at a time. The remote access capability allows computers at remote sites, not connected to the relay broadcast network, to connect and listen in on the data stream captured by the relay machine. This enables scientists to monitor their instruments without being present at the test site. In this mode, the program uses a TCP/IP connection to send the relay machine a list of data types it wants to receive. An independent thread (subprogram) running in the relay program then directs the requested subset of the data to the TCP connection. If the remote machine or the network cannot handle the data rate, the relay program will drop packets without affecting other functions of the relay machine.

4.3. Real Time Science Data Display

The primary output interface of the data capture program is to the FAST data analysis program, SDT. SDT is a science display software package, developed at Berkeley initially for Geotail and upgraded for several other projects including CRRES, Polar and FAST. SDT presents a menu driven interface to the user that allows rapid selection of data quantities to be plotted, and quick changes to the plot attributes (plot limits, line or points, integration limits, labels) and display attributes (plot selection, zoom, pan, plots/page, print, color table, etc.). SDT can exhibit time-series (multi-line) and spectrogram data, can run as an off-line 'batch' program, and can be used to feed data into IDL programs. It can be used to view data either in post-orbit data analysis mode, or in real-time mode for prelaunch testing, instrument turnon, and the FAST campaign. In real-time mode, SDT uses a circular RAM buffer to store the most recent data for capture and display on its screen in wrap around mode (oscilloscope). The circular buffer prevents memory overflow during long viewing periods, but prevents the user from viewing all the data with arbitrary plot limit changes. Since it is written using XWindows, it can also be used remotely (over the internet) as well as on stand alone machines.

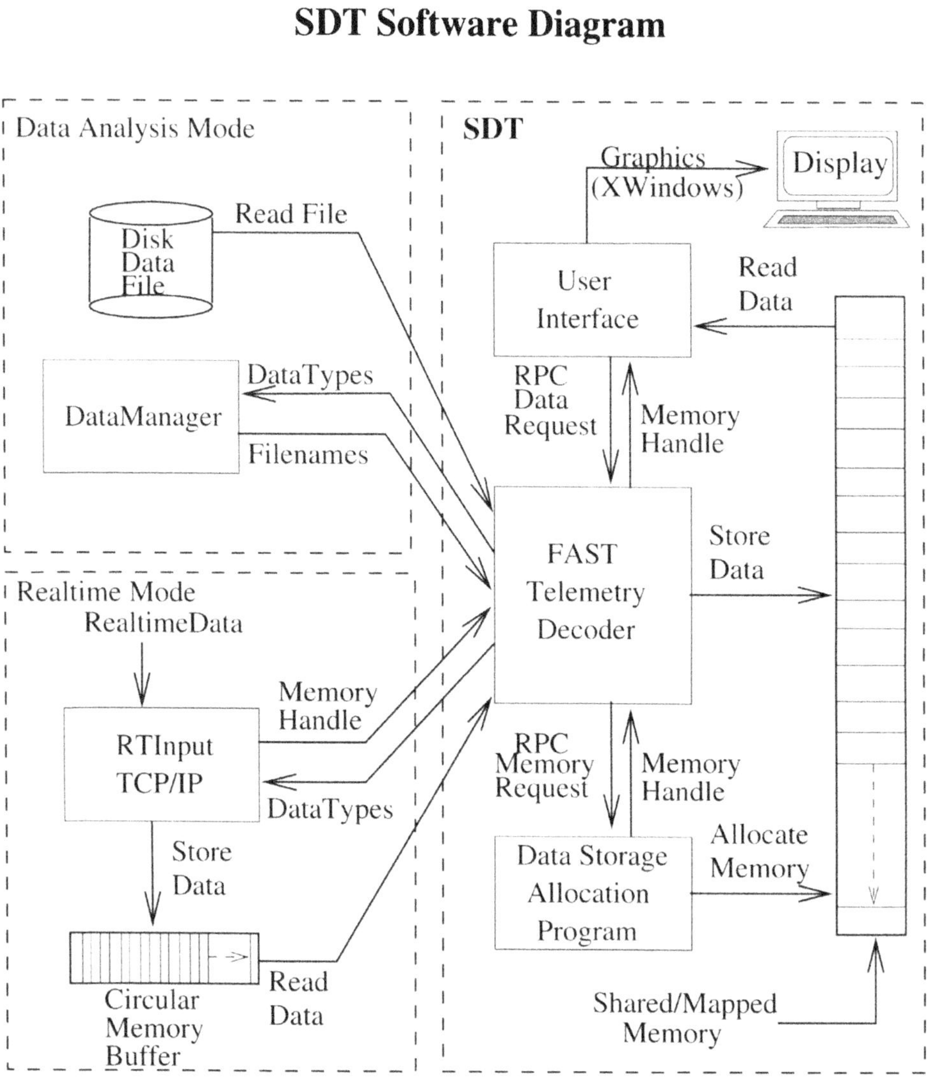

Figure 6. A block diagram of the real-time and data analysis program, SDT. The SDT program allows the same display software to be used for integration, launch and early orbit, and during the data analysis phase. SDT locates real-time data buffers through an interface with the RTInput program, or locates data files through an interface with the data manager program. SDT's telemetry decoder decommutates the data and stores it in memory in a format readily displayed by the user interface. The user interface allows rapid changes in data quantities to be plotted, and quick changes to the plot attributes (plot limits, line or points, integration limits, labels) and display attributes (plot selection, zoom, pan, plots/page, print, color table, etc.).

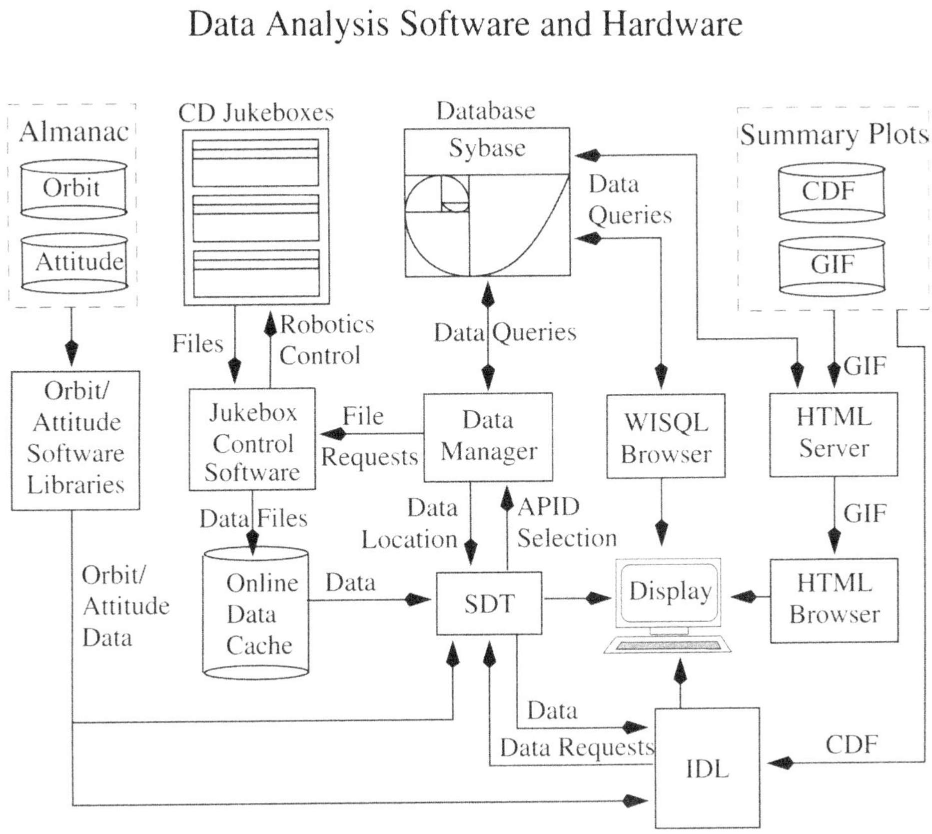

Figure 7. A block diagram of the data analysis software. The central programs are the data manager program and data analysis program, SDT. A typical data session would begin with the selection of science data plots within SDT followed by a call to the data manager to find a particular orbit or time. The data manager queries SDT to determine data types, accesses the database to determine file names, searches for the files on local hard disks, downloads those files not found from the data archive, and provides SDT with a list of file paths. SDT reads the files, decommutates the data placing it in memory, and plots the data on the fly. Once in memory, data can be further accessed through IDL. Browser programs are also available to provide database and key parameter searches.

A block diagram of SDT is shown in Figure 6. SDT consists of a user interface program, a data storage allocation program, and project specific programs, called ‘decommutators’ of which the FAST telemetry decoder is one, which decode the corresponding telemetry streams into data buffers which the user interface program can display. Use is made of multi-threading and various UNIX inter-process communication facilities such as Remote Procedure Calls, TCP Sockets, Shared-Memory, and Mapped-Memory files. For real-time FAST usage, data is read from the shared memory buffer which is written to by the data capture program (RTInput).

5. Data Analysis Software

FAST data are analyzed through a set of programs, both commercial and in-house, which allow the scientists quick and easy access to the data. Figure 7 illustrates the primary programs and hardware used by the FAST science team. Several browsing tools (WISQL Browser allows database queries, HTML browser allows GIF display of summary plots via FAST Web page, IDL displays CDF file summary data) are available to help select events for detailed investigation. Jukebox access software allows the user fast access to the entire data set. Once events are selected, the data can be loaded into RAM by the FAST data analysis tool, SDT, for detailed viewing as time series, or the data can be further accessed by IDL for detailed analysis. These programs and tools are outlined below.

5.1. Interface to On-line Data Storage

The FAST data is archived on CD-ROMs and stored on-line in jukeboxes (see Data Archive section). A commercial software package (Luminex Fire Series Software) interfaces to the jukeboxes making the system appear as a UNIX file system. Data analysis software requires that the files be downloaded to a local hard disk before being accessed by SDT or IDL. The database (Sybase) keeps information about which CD-ROM volume a data file resides on, and also keeps information about what data is stored in each file, such as timespan, instrument modes, and ephemeris data. Scientists can write SQL (Standard Query Language) queries to find interesting data, however, the primary tool for finding data is the data manager program described below. If specific data queries are required, the WISQL program provides a useful platform for initiating queries or browsing the database. (WISQL was downloaded from the web as part of 'sybtcl', an interface between Sybase and TCL developed by Tom Poindexter, http://www.nyx.net/~tpoindex/tcl.html#Sybtcl).

The data manager program provides the primary interface between the database, the jukeboxes, the data stored on-line on hard disk drives, and SDT. A scientist using SDT can select the data quantities he wants to analyze, and then select a timespan or orbit number in the data manager program. This program makes a database query for the files and CD-ROM volumes needed. The query can be further restricted by instrument mode, data type (application identifier or APID in Figure 7), and invariant latitude. The program obtains a list of data types from SDT before it makes the query, so that only the files needed by the scientist for the SDT session will be retrieved. When the data manager program receives the query results, it locates the files on hard disks, or allows the scientist to download the files from the jukebox. Upon completion of file downloading, the program sends SDT a list of file locations, and SDT decodes the data. Each scientist has a personal disk drive to which data can be downloaded and the data manager program keeps track of data files stored on all disks on the local network. The intermediate buffering of data files to the user's local hard disks improves the response time of the analysis programs for any data accessed more than once by the entire group (as is typical for interesting data), but also prevents any single user from overwhelming the system by requiring that the individual remain within his or her disk space limits.

Co-Investigators can use the data manager program at remote sites to download data they need from the jukeboxes at Berkeley via the internet. They can also use it to access their own jukebox, which could be loaded with a subset of the FAST data set. In both cases, they access the Berkeley database using Sybase client software bundled with the data manager program, so that it is not necessary to purchase and maintain Sybase for sites other than at Berkeley.

5.2. Science Data Display

Once files are located by the data manager program, SDT loads the data for display. Operation of SDT in data analysis mode is much like its operation in real-time mode, except that the data set is loaded into a linear memory buffer rather than the real-time circular buffer. A 'fastcount file', generated as part of the archive process and stored with the data, is used to determine memory requirements. This use of memory makes the program's response to changes in plot parameters quite fast, but requires the user's machine to have a large amount of RAM (> 100 MBytes) to view most data sets. When the data are available on local hard disk, SDT makes an excellent browsing tool. However, a few minute wait is initially required (depending upon the data volume) if downloading from the jukebox.

SDT can also be used as the first stage of data processing. Its 'batch' mode is used to create summary plots of housekeeping and basic instrument operation. For more advanced analysis, SDT can make data available to an IDL Batch session or produce disk files, which can then be read into other programming environments. SDT also has the ability to perform advanced analysis internally through the development of 'Science Modules'. These are subprograms that attach to SDT for the purpose of creating new data quantities (such as despun electric and magnetic field data) from one or more of the raw data quantities. Science modules must be written in C (or C++) and SDT has the appropriate interfaces and libraries to allow the addition of these programs.

In addition to viewing data using SDT, scientists require a data analysis tool that allows them to easily access and manipulate the raw data. To permit this, an interface between data stored by SDT and IDL was developed. The SDT to IDL interface consists of a library of C and C++ routines, and a group of IDL routines which extract the various generic SDT data types. These are used as the building blocks for data specific access routines which exist for each of the FAST data types. Once the data is extracted to IDL, the scientist is free to manipulate the data in a simple environment. The IDL programs are used in a command line environment rather than a menu-driven, point and click, program. Menu-driven programs are, in general, limiting since they require the developer to foresee any manipulation the user will consider. Since this is an impossible task, FAST scientists operating in the IDL environment are expected to learn the general data analysis tool routines. In practice this method can be much faster than menu driven programs since the user can quickly develop a 'crib-sheet' of custom commands and use cut and paste to execute multiple commands with a few clicks of a mouse.

5.3. Data Viewing on the Web

The data analysis effort is also supported by the FAST web site (http://sprg.ssl.berkeley.edu/fast). This site was created to provide the science team quick, timely access to numerous information products generated during the FAST mission and to provide the worldwide space physics community access to science data and general information on the mission. In general this information ranges from specific operational items (stored command loads, ground station contact schedules, orbit information, spacecraft and instrument status, instrument modes, and FAST documentation) to FAST science summary data which can be viewed and downloaded by the user. The science summary data link presents the user with a form which allows him or her to build a query to search for specific summary data by date/time or orbit. The user can further define the query for data type and auroral region. The user is presented with links to GIF, Postscript, and CDF files matching the query information. The FAST FTP site also contains the up-to-date data analysis and computer system software releases available to the FAST science team and associated support documentation.

6. Conclusion

The Science Operations Center for the FAST mission was developed to process and quickly route FAST data to the scientists, and to serve as an on-line data archive center for the project. The SOC also serves as an interface between the FAST investigators and NASA ground operations, providing daily command loads to spacecraft that optimize the science data collection and supporting near real-time commanding during spacecraft testing. A low cost operations center was developed by relying on automated programs and the latest available hardware. Software was developed that provides a smooth transition from pre-launch ground testing to post-launch data analysis. Extensive data analysis tools are available that allow the scientists to quickly access the large volume of data expected over the mission life (>1 terabyte). It is hoped that the above description not only provides an introduction to the FAST mission for those scientists wishing to collaborate in the future, but also serves as a general outline on how to design a low cost, science operation center with fast, convenient data analysis tools.

Appendix

A/D	Analog to Digital
ALT	Altitude
APID	Application Identifier
CD-ROM	Compact Disc - Read Only Memory
CDAWeb	Coordinated Data Analysis Workshops Web site run by NSSDC
CDF	Common Data Format
CDHF	Central Data Handling Facility
CRRES	Combined Release and Radiation Effects Satellite
FAST	Fast Auroral Snapshot Explorer
FIFO	First In First Out
FTP	File Transfer Protocol
GEI	Geocentric Earth Intertial coordinate system
GIF	Graphics Interchange Format
GSE	Ground Station Equipment
HSBM	High Speed Burst Memory
HTML	Hyper Text Markup Language
IDL	Interactive Data Language software from Research Systems Inc.
IGRF	International Geomagnetic Reference Field
IGSE	Instrument Ground Station Equipment
ILAT	Invariant Latitude
ISTP	International Solar-Terrestrial Physics
KP	Key Parameter
MLT	Magnetic Local Time
NASA	National Aeronautics and Space Administration
NSSDC	National Space Science Data Center (NASA/GSFC Code 633 Facility)
POCC	Payload Operations and Control Center
PPS	Packet Processing System
RAM	Random Access Memory
RTInput	Real Time Input program developed by FAST science team
SCSI	Small Computer System Interface
SDT	Science Data Tool
SMEX	Small Explorer Missions
SOC	Science Operations Center
SQL	Structured Query Language
TCL	Tool Command Language
TCP/IP	Transmission Control Protocol / Internet Protocol
TEAMS	Time-of-flight Energy, Angle, Mass Spectrograph
TOTS	Transportable Orbital Tracking Station
UDP	User Datagram Protocol
UT	Universal Time
VC	Virtual Channel
WISQL	Windowed Interactive Structured Query Language

References

Carlson, C. W., McFadden, J. P., Turin, P., Curtis, D. W., Magoncelli, A.: 2001, 'The Electron and Ion Plasma Experiment for FAST', *Space Sci. Rev.* (this issue).

Elphic, R. C., Means, J. D., Snare, R. C., Strangeway, R. C., Kepko, L., and Ergun, R. E.: 2001, 'Magnetic Field Instruments for the Fast Auroral SnapshoT Explorer', *Space Sci. Rev.* (this issue).

Ergun, R. E., Carlson, C. W., McFadden, J. P., Mozer, F. S., Delory, G. T., Peria, W., Chaston, C. C., Temerin, M., Roth, I., Muschietti, L., Elphic, R., Strangeway, R., Pfaff, R., Cattell, C. A., Klumpar, D., Shelley, E., Peterson, W., Möbius, E., and Kistler, L.: 1998, 'FAST Satellite Observations of Large-Amplitude Solitary Structures', *Geophys. Res. Lett.* **25**, 2041–2044.

Ergun, R. E. Carlson, C. W. Mozer, F. S. Delory, G. T., Temerin, M., McFadden, J. P., Pankow, D., Abiad, R., Harvey, P., Wilkes, R., Primbsch, H., Elphic, R., Strangeway, R., Pfaff, R., and Cattell, C. A.: 2001, 'The FAST Satellite Electric Field and Magnetic Field Instrument', *Space Sci. Rev.* (this issue).

Harvey, P. R. Curtis, D. W. Heetderks, H. D., Pankow, D., Rauch-Leiba, J. M., Wittenbrock, S. K., and McFadden, J. P.: 2001, 'The FAST Spacecraft Instrument Data Processing Unit', *Space Sci. Rev.* (this issue).

Klumpar, D. M., Möbius, E., Kistler, L. M., Popecki, M., Hertzberg, E., Crocker, K., Granoff, M., Tang, L., Carlson, C. W., McFadden, J., Klecker, B., Eberl, F., Kunneth, E., Kastle, H., Ertl, M., Peterson, W. K., Shelley, E. G., and Hovestadt, D.: 2001, 'The Time-of-Flight Energy, Angle, Mass Spectrograph (TEAMS) Experiment for FAST', *Space Sci. Rev.* (this issue).

NASA Document No. 822-DEM-1724, System, Tracking, and Acquisition Handbook for the Transportable Orbital Tracking System (TOTS), 1996.

Shi, J., Mao, T., Chesney, J., and Speciale, N.: 1993, 'Fast Auroral SnapshoT Explorer (FAST) packet processing system', *Proc. Inter. Telemetering Conf.*, 449–459.

THE TIME-OF-FLIGHT ENERGY, ANGLE, MASS SPECTROGRAPH (TEAMS) EXPERIMENT FOR FAST

D. M. KLUMPAR[1], E. MÖBIUS[2], L. M. KISTLER[2], M. POPECKI[2], E. HERTZBERG[1], K. CROCKER[2], M. GRANOFF[2], LI TANG[2], C. W. CARLSON[3], J. McFADDEN[3], B. KLECKER[4], F. EBERL[4], E. KÜNNETH[4], H. KÄSTLE[4], M. ERTL[4], W. K. PETERSON[1], E. G. SHELLY[1] and D. HOVESTADT[4]

[1]*Lockheed Martin, Advanced Technology Center, Palo Alto, CA 94304, U.S.A.*

[2]*Institute for the Study of Earth, Oceans and Space, University of New Hampshire, Durham, NH 03824, U.S.A.*

[3]*Space Sciences Laboratory, University of California Berkeley, Berkeley, CA, U.S.A.*

[4]*Max-Planck-Institut für extraterrestrische Physik, Garching, Germany*

Abstract. The Time-of-flight Energy Angle Mass Spectrograph (TEAMS) is being flown on the FAST Small Explorer mission to measure the 3-dimensional distribution function of the major ion species present in the lower magnetosphere. The instrument is similar to time-of-flight plasma analyzer systems that have been designed and planned for flight as CODIF (COmposition and DIstribution Function analyzer) on the four European Space Agency Cluster-II spacecraft and, as ESIC (Equator-S Ion Composition instrument) on Equator-S. This instrument allows the 3-dimensional distribution functions of individual ion species to be determined within $\frac{1}{2}$ spin period (2.5 s). Two-dimensional distributions are measured in 80 ms. These capabilities are crucial for the study of selective energization processes in the auroral regions of the magnetosphere. The design, operational characteristics, and test and calibration results for this instrument are presented. The sensor consists of a toroidal top-hat electrostatic analyzer with instantaneous acceptance of ions over 360° in polar angle. After post-acceleration of the incoming ions by up to 25 kV, a time-of-flight mass spectrograph discriminates the individual species. It has been demonstrated through calibration that the instrument can easily separate H^+, He^{2+}, He^+, O^+ and, for energies after post-acceleration of > 20 keV, even O_2^+ molecules. On-board mass discrimination and the internal accumulation of several distinct data quantities combined with the spacecraft's flexible telemetry formatting allow for instrument data rates from 7.8 kb s^{-1} to 315 kb s^{-1} to be telemetered to ground through the FAST centralized Instrument Data Processor.

1. Introduction

The early suborbital rocket and satellite investigations of the auroral region in the 1960s determined that precipitating electrons were the dominant contributor to auroral luminosity. As more in-depth investigations of auroral processes were carried out it was learned that ionospheric ions play a significant role in auroral electrodynamic processes.

Mass spectrometry performed in the magnetosphere revealed that a significant and variable component of the plasma population was of ionospheric origin (Shel-

Space Science Reviews **98:** 197–219, 2001.

ley *et al.*, 1972; Young *et al.*, 1977). At high polar altitudes, above the auroral region, precipitating O^+ ions, necessarily of geogenic origin, were observed during a magnetic storm at energies in excess of 10 keV with intensities greater than the accompanying protons (Shelley *et al.*, 1972). Direct measurement of outgoing accelerated ions using mass spectrometers at high altitudes were subsequently confirmed by Shelley *et al.* (1976), and by Sharp *et al.* (1977); and with non-mass-resolving ion analyzers at low altitude (Klumpar, 1979), demonstrating the direct outflow of ionospheric ions. Physical processes were found to be active within and above the auroral ionosphere that result in the direct injection of ionospheric ions into the magnetosphere.

The broad objective of the Time-of-flight Energy Angle Mass Spectrograph (TEAMS) investigation on FAST is to directly characterize the three dimensional distributions of ions within these interaction regions, in order to understand the evolution of the ion distributions from cold ambient ionospheric plasma to hot accelerated magnetospheric particles and to discern the physical mechanisms that drive this evolution.

2. Instrument Heritage

Following the successful application of time-of-flight (TOF) mass-spectrographs in space pioneered with the AMPTE mission in 1984 (Gloeckler *et al.*, 1985; McEntire *et al.*, 1985; Möbius *et al.*, 1985) TOF instruments have become a standard tool in space plasma physics. TOF instruments have been flown in various configurations on missions, such as VIKING, GIOTTO, VEGA, PHOBOS, *Ulysses*, and *Cassini*. They are also aboard the GGS spacecraft. The wealth of the ion composition results from these missions has clearly demonstrated the excellent capabilities of this type of instrumentation in both the energy range of the bulk plasma as well as in the energy gap between about 30 keV and several hundred keV nucl^{-1}, which existed between the regimes of magnetic mass spectrographs and cosmic-ray telescopes.

To simultaneously determine the 3-D distribution function and the mass-composition of ions, the concept of the fast 3-D plasma instrument based on the top-hat design after Carlson *et al.* (1982) and first flown on AMPTE/ IRM (Paschmann *et al.*, 1985) has been combined with the TOF technique and post-acceleration. A suite of similar instruments were designed for the ESA/NASA Cluster mission (Rème *et al.*, 1993) as CODIF, for FAST as TEAMS, and for Equator-S as ESIC. For a description of this instrument family see Möbius *et al.* (1998b).

In this paper we describe the Time-of-Flight Energy Angle Mass Spectrograph (TEAMS) flown on the FAST Small Explorer Mission. The major achievement of this instrument will be to allow the accumulation of full 3-dimensional velocity distribution functions for four major species (e.g., H^+, He^{2+}, He^+ and O^+) within $\frac{1}{2}$ spin period of the FAST spacecraft. The instantaneous coverage of a two-dimensional cut of the distribution function is possible within one energy sweep whose typical duration is 80–160 ms. This capability will significantly enhance our knowledge of the dynamics, heating and acceleration of multi-ion plasmas that are typical of the auroral region.

3. Instrument Description

The instrument combines the selection of incoming ions according to their energy per charge by electrostatic deflection in a toroidal analyzer with post-acceleration by up to 25 keV e^{-1} and subsequent time-of-flight analysis. A top view and a cross-sectional view of the sensor showing the basic principles of operation are presented in Figure 1(a). The instrument is mounted with its axis of symmetry perpendicular to the spacecraft spin axis. The front end of the sensor protrudes beyond the spacecraft surface so that its 360° aperture has a free field of view.

The instrument design and orientation complement the electrostatic analyzer (ESA) instrument package on FAST (Carlson *et al.*, 1998, 2001) in two ways. First, TEAMS provides mass resolution capability, while the ESAs measure total ion flux without resolving mass, but with high time resolution. Second, TEAMS is oriented with the spin axis in the plane of its viewing field, allowing the full 3-dimensional velocity distribution of all ions to be measured in half a spin period. The FAST spin period is maintained at 5 s. TEAMS is the only fully 3-D particle sensor on FAST. The ESAs are oriented with their viewing field perpendicular to the spin axis and thus cover the full pitch-angle range at any instant. In this way the highest time resolution can be achieved.

3.1. Sensor

The electrostatic analyzer (ESA) is a rotationally symmetric toroidal top-hat with a uniform response over 360° of polar angle. Toroidally symmetric electrostatic analyzers carry the advantage of an extended focal length in the polar direction as compared to spherical analyzers, allowing them to be matched to a following mass-resolving analyzer. Toroidal geometries have been been developed and used for space plasma analyzers in the past (e.g., Ghielmetti and Young, 1987; Young *et al.*, 1988; Rème, 1993; Shelley *et al.*, 1995). As illustrated in Figures 1(a) and 1(b), the analyzer consists of an inner toroid, to which a variable negative potential is applied, an outer toroid with a cut-out at the top, and a top-cap lifted above the outer toroid. The analyzer carries a toroidal plate spacing of 4 mm. The inner and outer toroid radii are 61 mm and 65 mm, respectively, about a center of curvature displaced 12.1 mm from the rotational symmetry axis. The inner toroidal sections are bridged by a 17.9° hemispherical section of radius 100.4 mm centered on a point 37.5 mm below the center of curvature of the toroid. Along the centerline the top cap is elevated 12.73 mm above the inner spherical section. Both

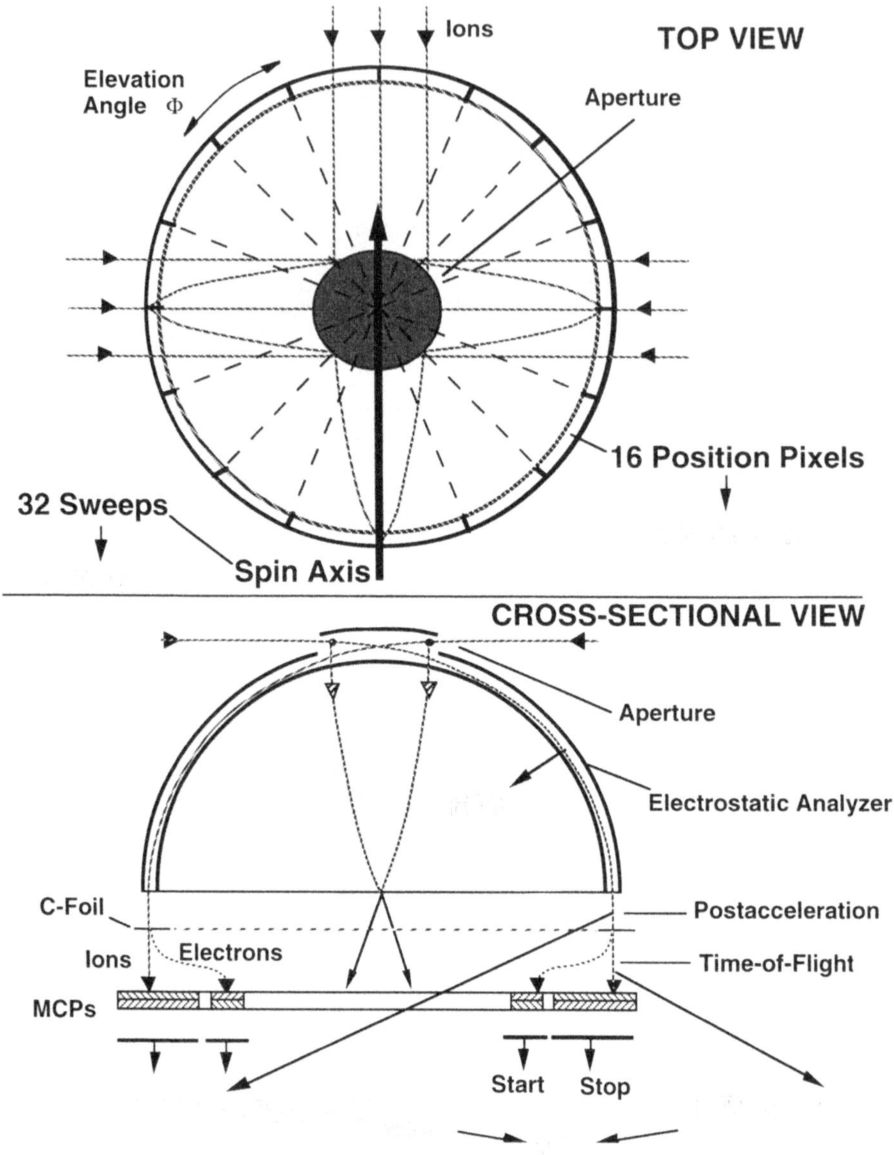

Figure 1. (a) Schematic top and cross-sectional views of sensor showing principles of operation. (b) TEAMS ion optics elements in cut-away view showing potential surfaces.

the outer toroid and the top-cap are held at ground potential, thereby exposing no high voltage to the external plasma. The outer toroid is scalloped to minimize the transparency to scattered ions and to reduce stray solar UV radiation. For the same reason the toroid surfaces are covered with a gold black coating. A beam of parallel ion trajectories entering the aperture is focused to a certain location at the exit plane of the analyzer. This location determines the incident polar angle of the

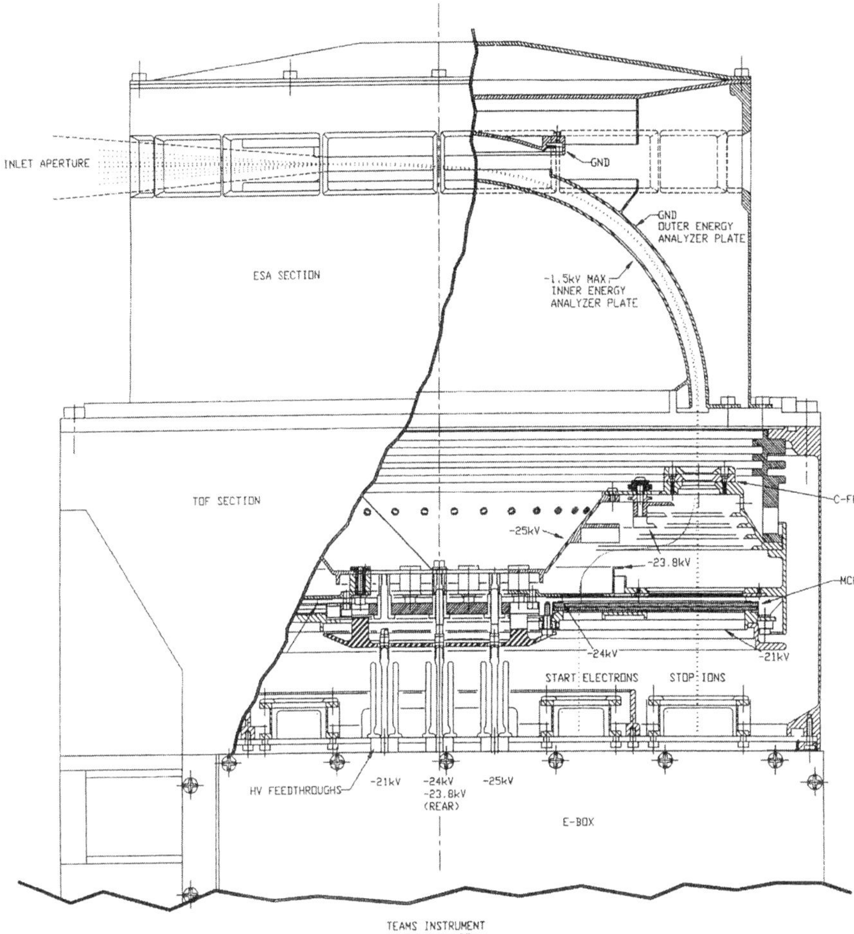

Figure 1. Continued.

ions. With an applied cross-plate voltage of 0.12 V to 1500 V (varied with 192 logarithmically spaced steps) incident ions with energies per charge from 1 eV e^{-1} to 12000 eV e^{-1} are selected. The analyzer has an intrinsic energy resolution of $\Delta E/E \sim 0.15$. It is surrounded by a cylindrical collimator which serves to define the acceptance angles. The full polar angle of the analyzer is divided into 16 pixels of 22.5° each. The full energy sweep is performed 32 times per spin (or 64 times per spin in high speed mode). Thus a two dimensional cut through the distribution function in polar angle with 11.25° (or 5.625°) sample spacing in azimuthal angle is obtained every 1/32 (or 1/64) of a spin period. The geometric factor for one individual pixel is $A\Delta E/E\Delta\theta\pi/8 \sim 1.5 \times 10^{-3}$ cm^2 sr, where A denotes the aperture area and $\Delta\theta$ the azimuthal width of the analyzer. This value takes into account the transmissivity of all grids and support structures in the sensor, but not the detection efficiency which varies with species and energy and will be discussed later.

Behind the energy per charge analyzer the ions are accelerated by a selectable post-acceleration voltage of -15 to -25 kV, such that the ions have at least an energy of 15 keV e^{-1} before entering the TOF-section. The energy per charge E/Q as selected by the electrostatic analyzer plus the energy eU_{ACC} gained by post-acceleration are combined with the measured time-of-flight τ through the length d of the time-of-flight (TOF) unit to give the mass per charge M/Q of the ion according to:

$$M/Q = 2(E/Q + eU_{\mathrm{ACC}} - E_{\mathrm{loss}})/(d/\tau)^2.$$

The quantity E_{loss} represents the effect of energy loss in the thin carbon foil ($\sim$ 3 μg cm^2) at the entry of the TOF section and depends on particle species and incident energy. The flight path of the ions is defined by the 3 cm distance between the carbon foil at the entrance and the surface of the stop microchannel plate (MCP). The start signal is provided by secondary electrons, that are emitted from the carbon foil during the passage of the ions. The electrons are accelerated and deflected onto the start portion of the MCP by the appropriate potential configuration (e.g., Figure 1(b)). The secondary electrons also provide the position information for the angular sectoring. To allow good angular resolution, a toroidal geometry has been chosen for the analyzer which pushes its focal point close to the carbon foil plane. The carbon foil is made up of 22.5° sectors, separated by narrow metal strips. The electron optics are designed to strongly focus secondary electrons originating at a foil onto the corresponding MCP start sector, using the fringe fields of the radial foil support structures and radial fin separators between the sectors. The MCP assemblies are ring shaped quadrants with radii of 3 by 9 cm (inner and outer radii, respectively) which serve both the start and stop signals.

The electron clouds exiting the MCPs are accelerated by the negative postacceleration potential to the signal pickup assemblies which are at ground potential. The signal assemblies consist of two sets of thin wire grids with $\approx$ 85% transmission at a distance of 10 mm in front of a segmented signal plate. The timing information is taken from the grids, separately for two 180° segments. The position of the incoming ions is taken from sixteen 22.5° signal plate pixels behind the start section of the MCPs complemented by four 90° pixels behind their stop section. In this way the timing and position signals are electrically separated already at the source of the signal.

3.2. Sensor Electronics

The sensor electronics consists of two time-to-amplitude converters (TACs) to measure the time-of-flight of the ions separately for two halves of the TOF section, sixteen position discriminators at the start and four position discriminators at the stop section of the MCPs, as well as event selection logic. A schematic block diagram is shown in Figure 2. A valid ion event requires the basic coincidence of a start and stop signal as well as the simultaneous registration of a start and stop position signal in the matching quadrant. In addition, only one start position at a time may be triggered. This configuration constitutes the normal operation mode. Via telecommand the requirements can be loosened by dropping the quadrant matching requirement. In addition the raw pulse height data can be configured to include events with multiple starts, mainly for diagnostic purposes. Each individual ion event is pulse height analyzed according to its time-of-flight, incidence in azimuthal (given by the spacecraft spin) and polar angle (given by the start position), and the energy per charge (given by the actual deflection voltage). In addition, counters accumulate rates individually for each start and position signal as well as for all time-of-flight measurements, all events with multiple start positions, events with only a single start position, and events that fall below a preset time-of-flight.

The postacceleration voltage is provided by a DC-DC converter with a 12 stage multiplier. The voltage can be set to voltages up to -25 kV via telecommand. The MCP bias voltage is supplied by a converter with a similar design, whose high voltage multiplier stages are referenced to the second-last stage of the postacceleration multiplier. The MCP voltage can be commanded up to 4 kV, thus providing adequate margin to compensate for potential gain degradation of the MCPs during the mission. The normal operating voltage is about 3 kV. The high voltage circuitry is protected by a current-sensitive circuit that turns the corresponding supply off upon detection of a discharge. The voltages and currents of all high voltage supplies are monitored by housekeeping ADCs along with the internal supply voltages of the sensor electronics and temperatures in the sensor, analog and digital electronics.

A detailed discussion of several critical design aspects of this instrument may be found in Möbius *et al.* (1998b).

4. TEAMS Interface to FAST Instrument Data Processing Unit

The instrument includes a hardware signal processing unit that compresses the original pulse height information and handles the sensor control functions and housekeeping data collection.

4.1. Data compression

To compress the data, look-up tables are stored on board which give the mass/charge bin, given the time-of-flight and energy/charge and the solid angle bin, given the input polar angle and azimuthal (spin) angle. Based on a comparison with these look-up tables each valid event is accumulated into incrementing memories; one with 48 energy bins, 64 solid angle bins, and 4 mass bins, and a separate memory with 16 energy, 16 solid angle, and 64 mass bins. While the first accumulator is designed to support detailed high time resolution studies of the velocity distributions

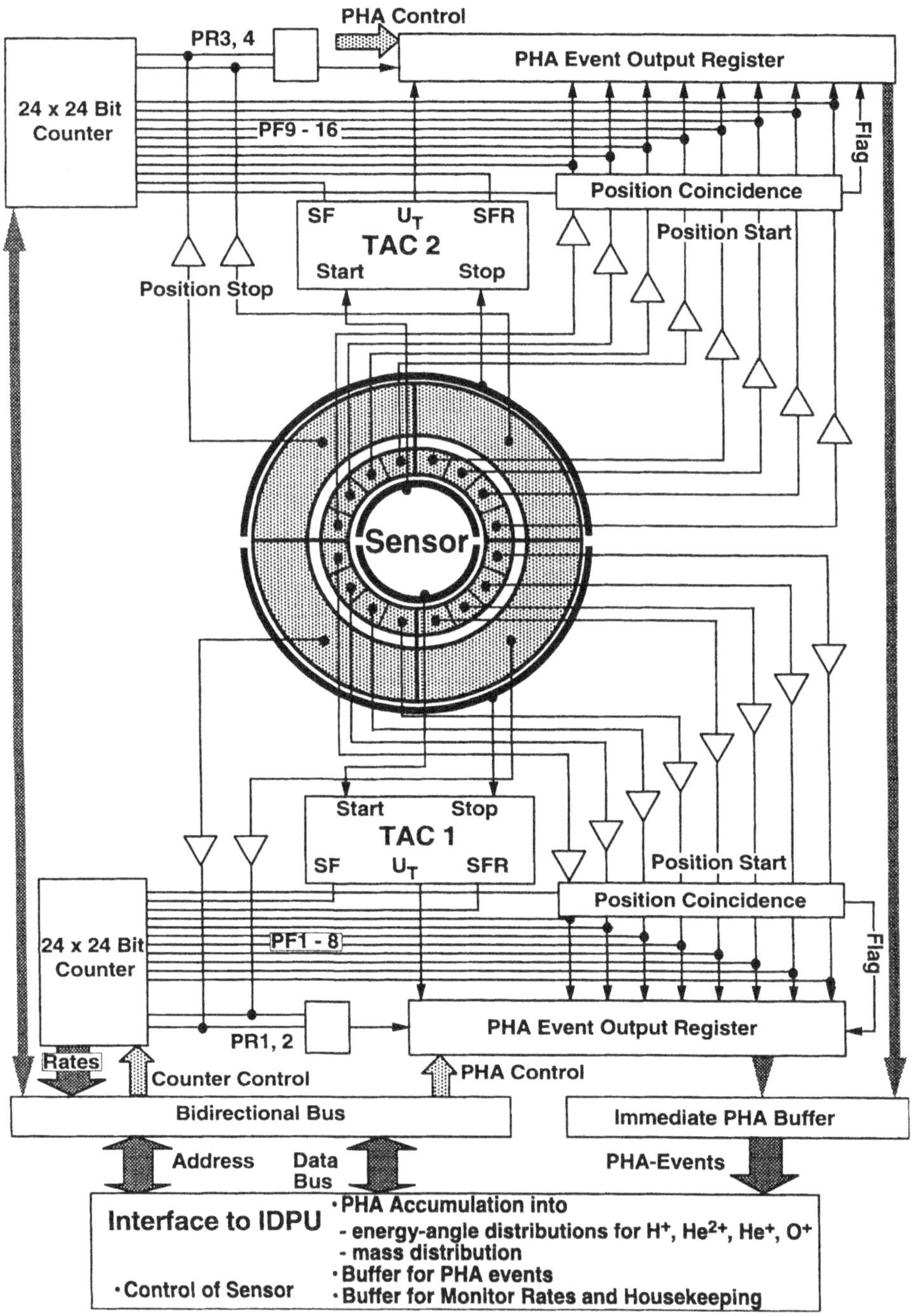

Figure 2. Sensor electronics block diagram.

of the 4 major species, the second one allows a more detailed composition study with longer accumulation times. In parallel, the events are also accumulated into a 4 species burst accumulator with the full resolution in azimuthal and elevation angle, whose contents is transmitted in conjunction with other high speed data products of the payload upon specially set trigger conditions. A separate memory allocation is set aside to accumulate high time resolution data for four species from the two pairs of angle sectors that lie adjacent to the spin axis. The resulting data products: survey, hi-mass, burst and pole data are described in more detail below. Because the mass/charge algorithm depends on the total energy of the ion (incoming energy plus post-acceleration), a new mass table must be loaded when the post-acceleration voltage is changed. The ability to upload new tables also allows flexibility in choosing the 4 species used in the survey, burst and pole data products.

In addition, a sample of raw pulse height events, monitor rates, and status data are taken and stored in a separate memory block. The memory blocks are read out by the centralized Instrument Data Processing Unit (IDPU) in 1 kB blocks. Rotating buffering of the memory assures the integrity of the data during readout.

4.2. Instrument Command and Control

The interface board also contains the command registers and high voltage control circuits. Both the postacceleration and MCP supplies are controlled by 8 bit DACs via telecommand. The sweep supply for the analyzer is controlled by a 16 bit DAC whose LSB is ignored. The sweep is performed from a sweep table in the ACTEL logic synchronized with the instrument clock and the spin phase. All HV supplies require an enable command in addition to a setting > 0.

When the payload is not taking data, TEAMS is placed in a low-power standby mode with only the interface electronics and the postacceleration supply powered. The instrument is brought into its fully operational condition by turning on in sequence the accumulator electronics, the sensor electronics and ramping up the MCP bias voltage to its operational value. The two time-of-flight circuits and related position amplifiers can be powered separately.

The upload of mass and angle look-up tables requires the instrument in a condition with only the digital accumulator electronics on. Simultaneous sensor signals would disrupt the memory write operations.

An Inflight Calibration circuit allows the stimulation of both the time-to-amplitude converters and the position amplifiers with simulated events, whose time-of-flight can be varied by a DAC. A predefined calibration sequence will be run every few weeks over the course of the mission to monitor the performance of the analog and digital electronics independently of the sensor performance.

5. Data Products and Instrument Modes

The TEAMS instrument produces five science data products. Two additional data products (monitor rates and status) provide instrument operational information at a basic level. Internally all data products are being produced and accumulated continuously. The payload telemetry mode determines which one or which combination of data products are read by the FAST IDPU and placed into the FAST data stream. In addition rates data are always telemetered, as an engineering monitor. The five science data products are: burst, survey, pole, high mass resolution (hi-mass), and PHA as described below:

5.1. Data products

5.1.1. *Burst*

Burst data is full resolution data for four mass groups (normally H^+, O^+, He^+, and He^{++}). For each step of an energy sweep (48 major E-steps) data is collected from all sixteen position pixels. The 32 (or 64) azimuthal look directions are accumulated consecutively over a spin period. These data are not compressed. Burst data is read into the IDPU irrespective of TEAMS telemetry mode. Its presence in the telemetered data stream depends upon spacecraft telemetry mode. Burst data are contained in application identifier (APID) 1047 (hex 417) in the FAST telemetry stream.

5.1.2. *Survey*

Survey data is basically high resolution energy and angle data consisting of 4 mass groups, 48 energies, and 64 solid angle segments. The 4 mass groups selected are H^+, O^+, He^+, and He^{++}. The 64 solid angle segments (organized by spin phase) consist of 16 equatorial segments (combining the pixels that map into the same solid angle element after $\frac{1}{2}$ spin, PF4+PF13 and PF5+PF12, respectively); 8 lower-mid-latitude, 4 upper-mid-latitude, and 4 polar segments in each hemisphere (see Figure 3(a)). The equatorial solid angle segments contain 2 (4) samples at each energy in the 32 (64) sweep/spin mode. Polar solid angle segments contain 8 and 16 samples at each energy for 32 and 64 sweep per spin modes, respectively. The full angular range is covered in half a spin but the actual time resolution of the measurement depends upon the telemetry mode. In the highest telemetry rate modes H^+ and O^+ survey data come out every half spin. In lowest telemetry rate mode these data are accumulated for 4 spins. In every mode He^+ and He^{++} are accumulated twice as long as H^+ and O^+ before being read out. The data are compressed from two bytes to one byte in the IDPU. Survey data are contained in APID 1027 (hex 403).

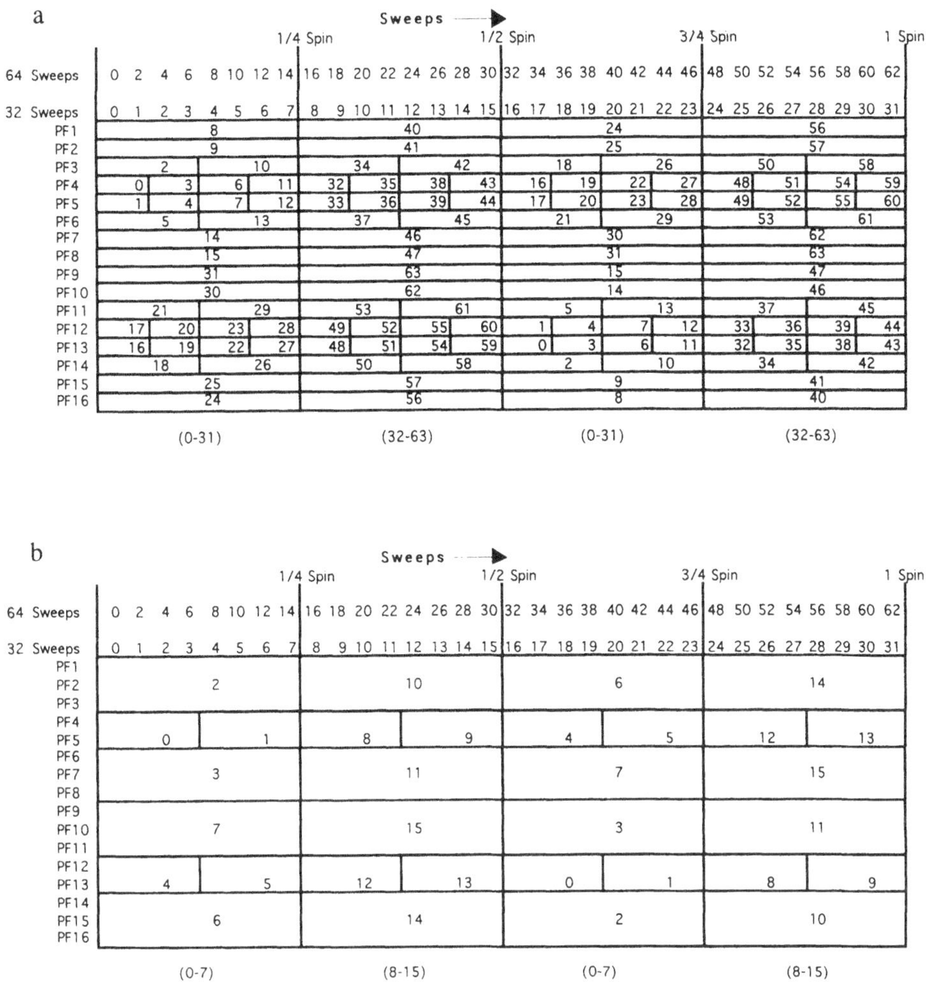

Figure 3. (a) Solid angle map for Survey data. (b) Solid angle map for High Mass data.

5.1.3. *Pole Data*

This science data quantity consists of high time resolution data along the spin axis. Because the spin axis is oriented perpendicular to the magnetic field in the key operation region of the satellite, 90° pitch-angle particles are collected in the Pole data. Data is accumulated from each of the two pairs of angle sectors that straddle the spin axis (PF1 and PF16 receive ions moving along the spin axis direction and PF8 and PF9 receive ions moving in the negative z-axis direction). Only the upper 32 energy steps are used for this data product (26 eV to 12 keV). The same four standard mass groups are accumulated, however the selected telemetry mode determines which mass groups are placed into the telemetry stream (see Table II). The data are accumulated for 8 sweeps and therefore are read out every $\frac{1}{4}$ spin for 32 sweep/spin modes and every $\frac{1}{8}$th spin for 64 sweep/spin modes. Pole data are contained in APID 1029 (hex 405).

TABLE I

PHA events data structure

Name	Byte#	Bits	Description
TOF	1	0–7	Time-of-flight channel number (0–255)
Start position	2	0–7	One bit for each start pixel
Energy Step	3	0–7	Microstep number (0–191)
Spin Angle	4	0–5	Angle 0–63
Section	4	6	Instrument half (0=>PF1–8(HS); 1=>PF9–16(LS))
Stop Position	4	7	1 if one stop pixel triggered, 0 if both triggered

5.1.4. *High Mass Resolution Data (hi-mass)*

The primary purpose of the high mass resolution science data quantity is to collect mass spectra over the full mass response of the spectrograph at relatively low time and energy resolution. Data are accumulated into 64 mass, 16 energy, and 16 solid angle segments. This array is read out by the IDPU every 16 s/c spins (80 s) for telemetry modes 0–3; every 8 s/c spins (40 s) for modes 4 and 5; and every 4 s/c spins (20 s) for modes 6 and 7. The mass bins cover from 0.6 to 60 AMU, in 64 logarithmically spaced steps. Each of the 16 energy bins is the sum of 3 adjacent major energy steps. The 16 solid angle segments cover the unit sphere as follows: eight azimuthal solid angle segments cover the equatorial band between ±22.5° (containing measurements from instrument pixels PF4, PF5, PF12, and PF13), and four azimuthal segments each cover the spin and antispin quadrants for polar angles from 22.5° to 90° (PF1-PF3 and PF14-PF16) and −22.5° to −90° (PF6-PF11) respectively (see Figure 3(b)). Hi-mass data are contained in APID 1028 (hex 404).

5.1.5. *PHA Events Data*

This science data product contains detailed information on a sample of individual pulse height events as collected by the instrument. Besides providing full resolution data, the main purpose of this product is to monitor the resolution and background performance of the sensor. This is necessary to keep the mass selection criteria used for the science data products optimized throughout the mission. The number of events collected per spin depends on the TEAMS mode, and varies from 128 to 2048 events spin^{-1}. A threshold can be set to eliminate events at low times-of-flight (e.g., to eliminate protons in order to concentrate on less abundant species). The events thrown out are counted and telemetered. Each event uses 32 bits and contains the following information shown in Table I.

A scheme has been devised to ensure that accumulated PHA events contain a sampling of different angles and energy steps. During each spin, one energy step is selected, and *N* events are collected for that step, for each sweep in the spin. During the next spin, the selected energy step number is incremented by 4. Thus after 16 spins, the full energy range has been covered. If the rates are low, so that N events are not collected in the selected step, events will continue to accumulate in subsequent steps. The number N varies with mode from 4 to 32. PHA data are contained in APID 1031 (hex 407).

5.1.6. *Monitor Rates Data*

This data quantity provides insight into the sensor efficiency and the operation of the analog and event selection electronics on a continuous basis. All spacecraft telemetry modes contain this product. These data are used to monitor a sampling of events as they occur on each start pixel, each stop quadrant, and to monitor valid and non-valid events. Singles rates provide a measure of the instrument background due to penetrating radiation. During nominal operation the ratios of the coincidence and singles rates data provide a measure of the instrument start-stop efficiencies and of microchannel plate gains. In in-flight calibration mode they provide a health check of the event selection logic. As with the PHA data, only one energy step is measured per spin, and the selected energy step increments by four at the start of each spin. The subcommutation scheme is identical to that used for the PHA data. Rates data are contained in APID 1030 (hex 406).

5.1.7. *Status Data*

This data product contains an image of the actual command settings of the sensor. Status data are contained in APID 1059 (hex 423).

5.1.8. *Housekeeping Data*

The housekeeping is comprised of all high voltage supply voltages and supply currents, the internal + and − 5 and 12 V voltages, as well as the temperature at the MCPs, in the time-of-flight electronics, on the postacceleration HV supply and in the digital electronics. These values are read by the IDPU, and come out in the IDPU telemetry data products.

5.2. Operating Modes

There are ten basic science telemetry modes producing instrument bit rates from 7.8 kbps to 315 kbps. Three additional engineering modes are defined. Table II shows the output frequency of each of the TEAMS science data products as a function of science mode. The mode nomenclature in the top line of Table II refers to two Slow Survey (SS) Modes (modes 0 and 1), six Fast Survey (FS) modes, and two Burst (Bst) Modes. When the spacecraft is in a 'Burst' mode, the burst data is telemetered in addition to the data from one of the Fast Survey modes.

TABLE II

TEAMS telemetry modes chart. Output frequency (in spins) of each data product versus telemetry mode

Mode name	SS-2	SS-1	FS-1	FS-1a	FS-2	FS-2a	FS-3	FS-3a	Bst-B	Bst-A
Mode number	(0)	(1)	(2)	(3)	(4)	(5)	(6)	(7)	(8)	(9)
# Sweeps/spin	32	32	64	32	64	32	64	32	64	32
Survey–H^+, O^+	2	4	2	2	1	1	0.5	0.5		
Survey–He^{++}, He^+	4	8	4	4	2	2	1	1		
Hi mass	16	16	16	16	8	8	4	4		
Pole–H^+, O^+	–		0.16	0.031	0.016	0.31	0.016	0.031		
Pole–He^+, He^{++}	–						0.016	0.031		
Burst									0.016	0.031
Rates	0.031	0.031	0.016	0.031	0.016	0.031	0.016	0.031		
PHA*	128	128	512	256	1024	512	2048	1024		
Status	1	1	1	1	1	1	1	1	1	1

*PHA is shown as # of events per spin.

6. Test and Calibration Results

A compilation of time-of-flight spectra for the key ion species at about 20 keV charge^{-1} total energy (incoming energy plus energy added by post-acceleration) is shown in Figure 4(a). This is the resolution achieved at low energies when the post-acceleration is operating near -20 kV. Because the major species in the magnetosphere are separated by factors of 2 in mass charge^{-1} a clear identification of these ions is possible. As can also be seen, ions with close masses such as N^+ and O^+ cannot be easily distinguished. The peak at mass per charge 2 AMU e^{-1} (He^{++} or H_2^+) occurs at time-of-flight values where the H^+ counts have dropped to 1% of their peak value. Therefore He^{++} abundances of 1% are difficult to distinguish, but higher abundances can be measured. Figure 4(b) shows an example of the time-of-flight spectrum measured during a pass over the aurora. The H^+ peak has been truncated at 1000 counts in order to better show the heavy ion peaks. The post-acceleration is operating at -19.7 kV at this time, and counts from energies up to 1 keV incoming have been included in the figure. The peaks are wider than in Figure 4(a) because the accelerator beams are monoenergetic while the flight peaks reflect both the acceptance energy width of the electrostatic analyzer, and the 1 kV band included in the figure to increase statistics. The O^+ flight peak is at higher times-of-flight because the calibration data point is for 21 keV total energy, while the flight O^+ is dominiated by ions close to 19.7 keV. It can be seen that the goal of separating the major species has been met with the present design.

While the response of an ESA remains essentially unchanged for different species and over the entire energy range, the detection efficiency of TOF assemblies changes with species and energy due to dependencies of the secondary electron production and the sensitivity of the MCPs. To reduce the anticipated variations the MCPs are operated in saturation of their gain. Yet there are still significant variations of the efficiency with species, energy, and position on the MCPs that may be as large as a factor of 2. Therefore, the sensor has been calibrated in a matrix of ion species, energies and MCP position. The final calibration was performed using the University of Bern ion beam test and calibration facility. A monoenergetic ion beam of known flux is directed at the instrument. The collimated beam fully illuminates the entrance aperture. The ratio of the instrument output to the known input is then used to determine the overall efficiency. Figure 5(a–c) shows the efficiencies for each of the three species, H^+, He^+, and O^+, as a function of energy in each of the 16 positions on the sensor. For He^+ and O^+, an average curve is also shown. This curve is based on measurements with the Cluster CODIF instrument, which covered a larger energy range. For these species, the average curve fit the individual curves well, except for an overall normalization factor. For H^+, a few of the positions were significantly different, so individual fits to the pixels were used for the final calibration factors. The efficiency for He^{++} is the same as the efficiency for He+ at the same total energy.

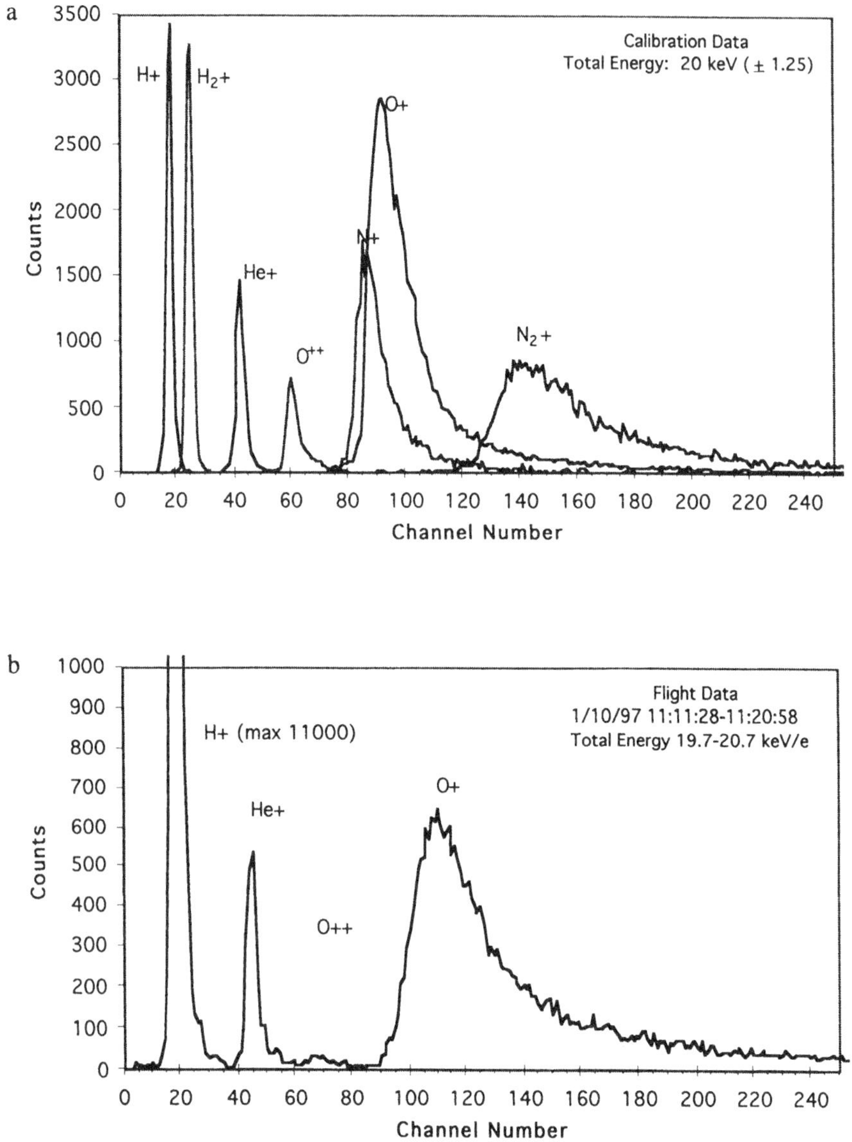

Figure 4. (a) A compilation of time-of-flight spectra for the key ion species at about 20 keV charge^{-1} total energy (incoming energy plus energy added by post-acceleration). This is the resolution achieved at low energies when the post-acceleration is operating near −20 kV. (b) A time-of-flight spectrum measured during a pass over the aurora. The H^+ peak has been truncated at 1000 counts in order to better show the heavy ion peaks. The post-acceleration is operating at −19.7 kV at this time, and counts from energies up to 1 keV incoming have been included in the figure.

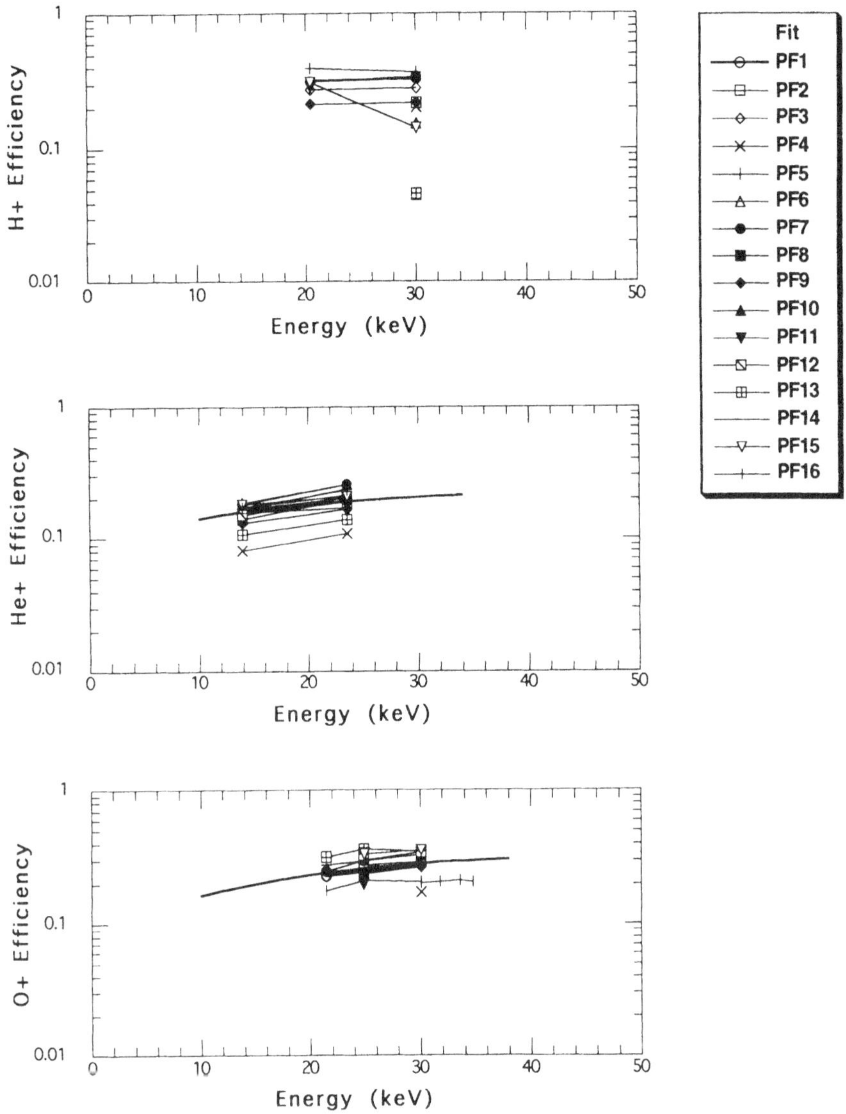

Figure 5. Total efficiencies for the species H^+, He^+, and O^+ as a function of energy.

Figure 6 shows the acceptance region of the sensor in electrostatic analyzer voltage and azimuthal angle (in the direction of the ESA deflection). The overall acceptance region is typical for this type of ESA. One deviation from the norm is the two-ridge structure, with the valley in between. Normally, just one ridge is observed. This feature is due to the fact that the incoming ion beam is focused by the post-acceleration field at the TOF entrance, and the resulting beam is so narrow that the carbon foil support grid structure becomes visible. The variation of the response as a function of ESA voltage (in a) and azimuthal angle (in b) is shown in Figure 7. In both cases the response has been integrated over azimuthal angle (in

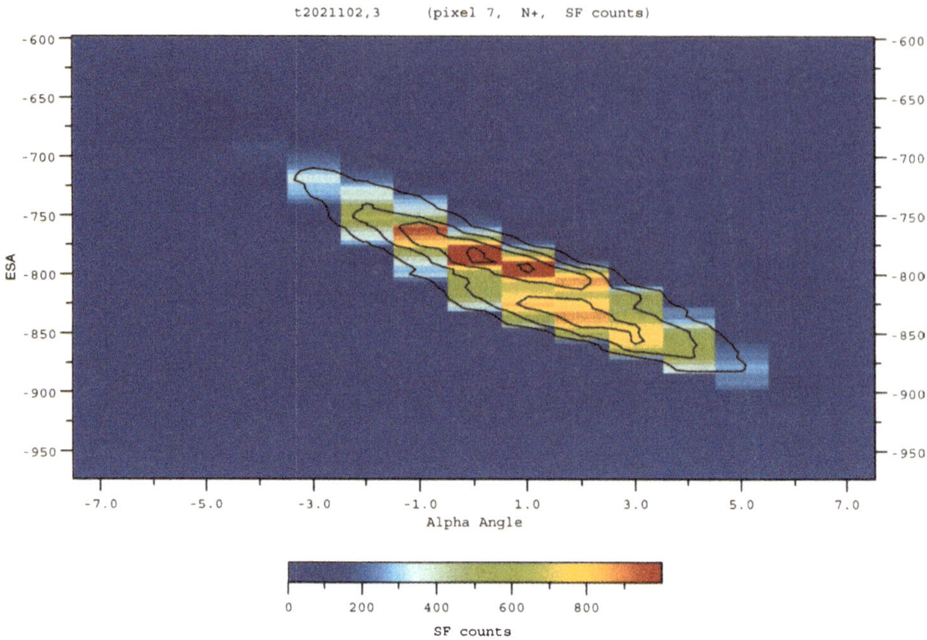

Figure 6. Example of the response of the sensor in energy and elevation angle.

a) and ESA voltage (in b), respectively. Except for the small minimum observed at the peak in a), the curves closely match the theoretical response of the ESA. The remaining effect of the carbon foil support grids is a reduction of the detected flux according to the transmission factor of the grids, which is included in the overall geometric factor.

7. In-Flight Performance

Figure 8 shows an overview of FAST plasma measurements including the TEAMS ion composition results during a typical auroral pass in the northern hemisphere for January 23, 1997, at about 4100 km in the premidnight magnetic local time sector. Shown for a 4-minute interval are from top to bottom: the electron ESA energy-time (E-t) spectrogram for 5–30 000 eV, averaged over all pitch-angles; the TEAMS H^+ E-t spectrogram (0–360° pitch angle), the TEAMS H^+ E-t spectrogram for the pitch angle subset 150° to 210°; O^+; and He^+ spin-averaged E-t spectrogram for 1–12 000 eV; and the pitch-angle vs time spectrogram for ions measured by the ion ESA without mass discrimination. The four ion composition panels are produced from the four angle sectors adjacent to the spin plane using the TEAMS survey data set. The Fast-Survey mode in use at this time produces half-spin resolution data for hydrogen and oxygen and one-spin resolution data for the helium ions.

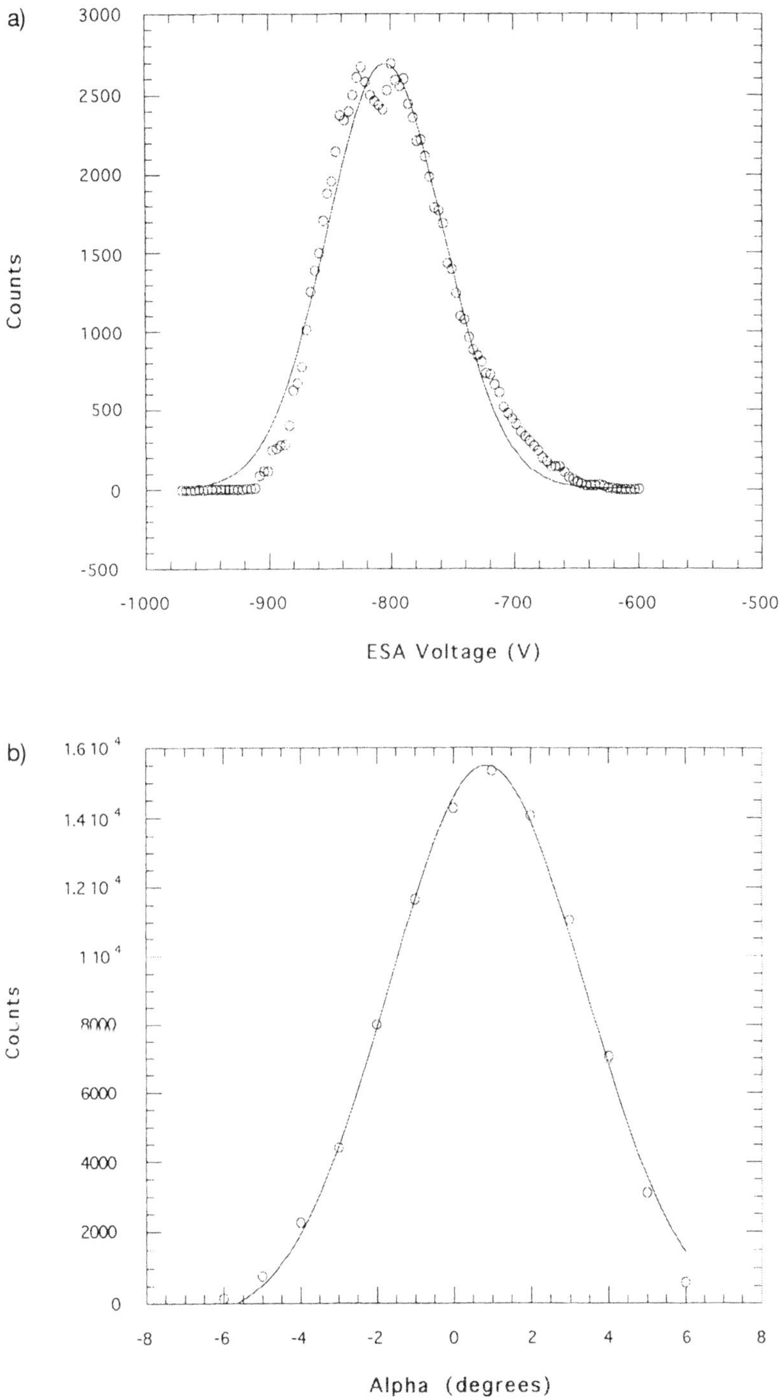

Figure 7. (a) Response of instrument integrated over azimuthal angle. (b) Response of instrument integrated over energy.

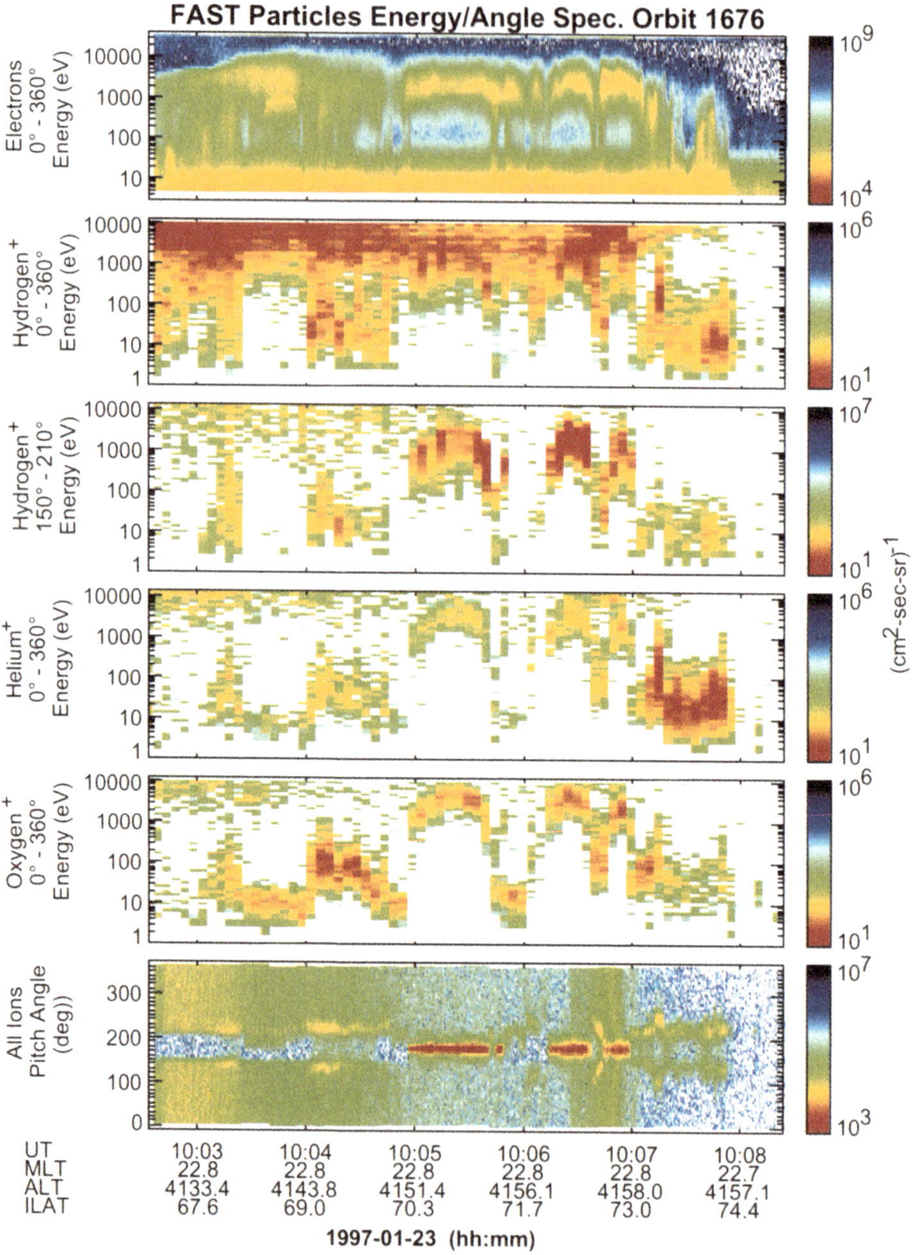

Figure 8. Overview of the FAST plasma measurements.

Four distinct upflowing ion beam events are present during the period shown, most clearly delineated by the intense flux covering a few degrees in pitch angle at 180° as illustrated by the red horizontal stripes in the lower panel beginning at 1004:55, 1005:50, 1006:10, and 1006:50 UT. The strong ion beams with energies up to several keV are observed along with the occurrence of electron cavities (indicated by the reduction of electron flux below 1000 eV in the top panel). The hydrogen in the upflowing beams is most clearly discerned by selecting only upward moving ions, as was done in the third panel, thereby excluding ions precipitating out of the magnetosphere. The precipitating ions are clearly present in panel two, where they have not been excluded, at energies above about 1 keV. The ion beams consist of H^+, He^+ and O^+ with varying relative abundances and with peak energies that systematically vary with mass. A study of these mass dependent ion beam variations is contained in Möbius *et al.* (1998a).

Also present in these data are ions whose pitch angle distributions peak near, or just above, the edge of the loss cone. These distributions, known as Transverseley Accelerated Ions, or 'Conics', are indicated in the bottom panel by the less distinct (and less intense) yellow shades centered about 40° away from 180° pitch angle beginning at 1004:00, 1006:05, 1006:40, and 1007 UT. These ions are also being accelerated out of the ionsophere, but through a distinctly different process than the one that acts to form the ion beams. As the E-t spectrograms show typical ion energies are lower in these distributions than in the ion beams.

Acknowledgements

The instrument team is indebted to the many unnamed individuals in the electronic and machine shops as well as administrative and support units, at all institutions involved in the hardware development and implementation of the instrument. The development, fabrication and test of the instrument is being supported by NASA under contract #NAS5-31283. Calibrations at the Bern calibration facility were supported by the Swiss National Science Foundation.

References

Carlson, C. W., McFadden, J. P., Turin, P., Curtis, D. W., and Magoncelli, A.: 2001, 'The Electron and Ion Plasma Instrument on FAST', *Space Sci. Rev.*, this issue.

Carlson, C. W. and McFadden, J. P.: 1998, 'Design and Application of Imaging Plasma Instruments', *Measurement Techniques in Space Plasmas*, American Geophysical Union, *Geophys. Monog.* **102**, 125–140.

Carlson, C. W.: 1992, 'The Fast Auroral Snapshot Explorer: June 9, 1992', *EOS* **73**, 249.

Carlson, C. W., Curtis, D. W., Paschmann, G., and Michael, W.: 1982, 'An Instrument for Rapidly Measuring Plasma Distribution Functions with High Resolution', *Adv. Space Res.* **2**, 67.

Ghielmetti, A. G. and Young, D. T.: 1987, 'A Double-Focusing Toroidal Mass Spectrograph for Energetic Plasmas, I, Theory', *Nucl. Instr. Meth.* **A258**, 297.

Gloeckler, G., Ipavich, F. M., Studemann, W., Wilken, B., Hamilton, D. C., Kremser, G., Hovestadt, D., Gliem, F., Lundgren, R. A., Rieck, W., Tums, E. O., Cain, J. C., MaSung, L. S., Weiss, W., and Winterhof, P.: 1985, 'The Charge-Energy-Mass Spectrometer for 0.3–300 keV/e Ions on the AMPTE/CCE', *IEEE Trans. Geosci. Remote Sens.* **GE-23**, 234–240.

Hamilton, D. C., Gloeckler, G., Ipavich, F. M., Lundgren, R. A., Sheldon, R. B., and Hovestadt, D.: 1990, 'New-High Resolution Electrostatic Ion Mass Analyzer Using Time-of-Flight', *Rev. Sci. Instrum.* **61**, 3104–3106.

Hovestadt, D. *et al.*: 1989, 'CELIAS – Charge, Element and Isotope Analysis System for SOHO', *The SOHO Mission*, **ESA SP-1104**, 69.

Klumpar, D. M.: 1979, 'Transversely Accelerated Ions: An Ionospheric Source of Hot Magnetospheric Ions', *J. Geophys. Res.* **84**, 4229.

McComas, D. J. and Nordholt, J. E.: 1990, 'New Approach to 3-D, High Sensitivity, High Mass Resolution Space Plasma Composition Measurements', *Rev. Sci. Instrum.* **61**, 3095–3097.

McEntire, R. W., Keath, E. P., Fort, D. E., Lui, A. T. Y., and Krimigis, S. M.: 1985, 'The Medium-Energy Particle Analyzer (MEPA) on the AMPTE CCE spacecraft', *IEEE Trans. Geosci. Remote Sens.* **GE-23**, 230–233.

Möbius, E., Tang, L., Kistler, L. M., Popecki, M., Lund, E. J., Klumpar, D. M., Peterson, W., Shelley, E. G., Klecker, B., Hovestadt, D., Carlson, C. W., Ergun, R., McFadden, J. P., Mozer, F., Temerin, M., Cattell, C., Elphic, R., Strangeway, R., and Pfaff, R.: 1998a, 'Species Dependent Energies in Upward Directed Ion Beams over Auroral Arcs as Observed with FAST TEAMS', *Geophys. Res. Lett.* **25**, 2029–2032.

Möbius, E., Kistler, L. M., Popecki, M. A., Crocker, K. N., Granoff, M., Jiang, Y., Satori, E., Ye, V., Rème, H., Sauvaud, J. A., Cros, A., Aoustin, C., Camus, T., Médale, J.-L., Rouzaud, J., Carlson, C. W., McFadden, J. P., Curtis, D., Heetdirks, H., Croyle, J., Ingraham, C., Klecker, B., Hovestadt, D., Ertl, M., Eberl, F., Kästle, H., Künneth, E., Laeverenz, P., Seidenschwang, E., Shelley, E. G., Klumpar, D. M., Hertzberg, E., Parks, G. K., McCarthy, M., Korth, A., Rosenbauer, H., Gräve, B., Eliasson, L., Olsen, S., Balsiger, H., Schwab, U., and Steinacher, M.: 1998b, 'The 3-D Plasma Distribution Function Analyzers with Time-of-Flight Mass Discrimination for CLUSTER, FAST, and Equator-S', *Measurement Techniques in Space Plasmas, American Geophysical Union, Geophys. Monog.* **102**, 243–248.

Möbius, E., Bochsler, P., Ghielmetti, A. G., Hamilton, D. C.: 1990, 'High Mass Resolution Isochronous Time-of-Flight Spectrograph for Three-Dimensional Space Plasma Measurements', *Rev. Sci. Instrum.* **61**, 3609–3612.

Möbius, E., Gloeckler, G., Hovestadt, D., Ipavich, F. M., Klecker, B., Scholer, M., Arbinger, H., Höfner, H., Künneth, E., Laeverenz, P., Luhn, A., Tums, E. O., and Waldleben, H.: 1985, 'The Time-of-Flight Spectrometer SULEICA for Ions of the Energy Range 5–270 keV/charge on the AMPTE/IRM', *IEEE Trans. Geosci. Remote Sens.* **GE-23**, 274–279.

Paschmann, G., Loidl, H., Obermayer, P., Ertl, M., Laborenz, R., Sckopke, N., Baumjohann, W., Carlson, C. W., and Curtis, D. W.: 1985, 'The Plasma Instrument for AMPTE/IRM', *IEEE Trans. Geosci. Remote Sens.* **GE-23**, 262–266.

Rème, H., Bosqued, J. A. Sauvaud, J. A., Cros, A., Dandouras, J., Aoustin, C., Martz, Ch., Médale, J. H., Rouzaud, J., Möbius, E. *et al.*: 1993, 'The Cluster Ion Spectrometry Experiment', **ESA SP-1159**, 133–162.

Sharp, R. D., Johnson, R. G., and Shelley, E. G.: 1977, 'Obseration of an Ion Acceleration Mechansim Producing Energetic (keV) Ions Primarily Normal to the Geomagnetic Field Direction', *J. Geophys. Res.* **82**, 3324–3328.

Shelley, E. G., Ghielmetti, A. G., Balsiger, H., Black, R. K., Bowles, J. A., Bowman, R. P., Bratschi, O., Burch, J. L., Carlson, C. W., Coker, A. J. *et al.*: 1995 'The Toroidal Imaging Mass-Angle Spectrograph (TIMAS) for the Polar Mission', *Space Sci. Rev.* **71**, 497–530.

Shelley, E. G., Sharp, R. D., and Johnson, R. G.: 1976, 'Satellite Observations of an Ionospheric Acceleration Mechanism', *Geophys. Res. Lett.* **3**, 654–657.

Shelley, E. G., Johnson, R. G., and Sharp, R. D.: 1972, 'Satellite Observations of Energetic Heavy Ions During a Geomagnetic Storm', *J. Geophys. Res.* **77**, 6104–6110.

Young, D. T., Bame, S. J., Thomsen, M. F., Martin, R. H., Burch, J. L., Marshall, J. A., and Reinhard, B.: 1988, '2-pi-Radian Field-of-View Toroidal Electrostatic Analyzer', *Rev. Sci. Instr* **59**, 743–751.

Young, D. T., Geiss, J., Balsiger, H., Eberhardt, P., Ghielmetti, A., and Rosenbauer, H.: 1977, 'Discovery of He^{2+} and O^{2+} Ions of Terrestrial Origin in the Outer Magnetosphere', *Geophys. Res. Lett.* **4**, 561–564.

GPSR Compliance

The European Union's (EU) General Product Safety Regulation (GPSR) is a set of rules that requires consumer products to be safe and our obligations to ensure this.

If you have any concerns about our products, you can contact us on ProductSafety@springernature.com

In case Publisher is established outside the EU, the EU authorized representative is:

Springer Nature Customer Service Center GmbH
Europaplatz 3
69115 Heidelberg, Germany

Batch number: 10614940

Printed by Printforce, the Netherlands